Synthesis Lectures on Mechanical Engineering

This series publishes short books in mechanical engineering (ME), the engineering branch that combines engineering, physics and mathematics principles with materials science to design, analyze, manufacture, and maintain mechanical systems. It involves the production and usage of heat and mechanical power for the design, production and operation of machines and tools. This series publishes within all areas of ME and follows the ASME technical division categories.

Carlos Alberto Dutra Fraga Filho

Reflective Boundary Conditions in SPH Fluid Dynamics Simulation

Two and Three-dimensional Validation and Applications

 Springer

Carlos Alberto Dutra Fraga Filho
Vitória, Brazil

ISSN 2573-3168   ISSN 2573-3176 (electronic)
Synthesis Lectures on Mechanical Engineering
ISBN 978-3-031-71581-5   ISBN 978-3-031-71582-2 (eBook)
https://doi.org/10.1007/978-3-031-71582-2

Thine eyes did see my substance, yet being unperfect; and in thy book all my members were written, which in continuance were fashioned, when as yet there was none of them.

Psalm 139:16 KJV

This book is dedicated to all the people (professors, students, researchers and collaborators) who have followed the implementation of reflective boundary conditions in particle simulation since the beginning of the author's research work. Furthermore, I dedicate it to God, who granted me the health and wisdom to produce it in times of scarce research and scientific dissemination support.

Preface

This book provides a synopsis of the efforts to implement and validate the physical reflective boundary technique coupled with Lagrangian particle modelling in continuum mechanics. In addition to the physics theory that justifies the use of the reflective technique in the continuous domain, it was essential to demonstrate that the implementation of this technique was feasible in the computational realm.

The characteristics of Lagrangian particle modelling, which provide the input data for the collision detection and response algorithm (the tool used to implement reflective boundary conditions computationally), are not the main focus of this work; however, the reader is provided with bibliographical references on particle modelling used in each case study in this book.

The first implementations in two-dimensional domains occurred about ten years ago. As the numerical results were validated, new implementations were carried out until the collision detection and response algorithm was possible to deal with collisions between particles and solid contours in three-dimensional domains.

This work's scientific contribution lies in compiling and synthesising the implementation's reflective boundary conditions and the validation stages for some relevant and well-known hydrodynamics and hydrostatics problems, which were historically solved using artificial boundary treatment techniques.

Vitória, Brazil
July 2024

Carlos Alberto Dutra Fraga Filho

Acknowledgements The author thanks Springer Nature for the opportunity to publish this book, which contains material of interest for students and researchers involved in computational simulation using Lagrangian Particle Methods and recent implementations of realistic physical boundary conditions.

I also wish to thank Profs. Alejandro Crespo and José Manuel Domínguez for kindly allowing me to use the content of Fig. 4.8.

Contents

About the Author

Carlos Alberto Dutra Fraga Filho is a Brazilian professor and researcher. He holds the following degrees: undergraduate in Mechanical Engineering (1998), Master of Science in Mechanical Engineering (2007), and Ph.D. in Environmental Engineering (2014). He has experience in the mechanical and environmental engineering fields, with an emphasis on Computational Fluid Dynamics and Transport Phenomena. Fraga Filho is a Journals' Reviewer: *Physics of Fluids* (American Institute of Physics), *Journal of the Brazilian Society of Mechanical Sciences and Engineering*, and *Advances in Water Resources*, and has two books—*Smoothed Particle Hydrodynamics: Fundamentals and Basic Applications in Continuum Mechanics* (https://doi.org/10.1007/978-3-030-00773-7) and *Dynamic Analysis of Composite Materials (FGM) via Finite Elements: Numerical Simulation and Experimental Validation* (ISBN-10: 6202043490)—and diverse scientific papers published.

Symbols

Roman Symbols

C_0	Initial position of the centre of mass
C_1	Position of the centre of mass at time $(t_0 + \Delta t)$
C_0'	Position of the centre of mass at the moment the particle has contact with the plane
C_{1E}	Position of the centre of mass at time instant $(t_0 + \Delta t)$, found using Euler's integration method
C_{1H}	Position of the centre of mass at time instant $(t_0 + \Delta t)$, found using the higher-order integration method
$(C_1)_N$	Coordinate of the particle's centre of mass perpendicular to the collision plane, in a motion disregarding the boundary, at the end of the numerical iteration
C_f	Position of the centre of mass obtained after the collision response (reflection)
$(C_f)_N$	Coordinate of the position of the centre of mass perpendicular to the collision plane after the collision response (reflection)
CF	Coefficient of friction
CR	Coefficient of restitution of kinetic energy
d	Distance between the particle's centre of mass and a plane
d_A	Positive distance between the particle's centre of mass and a plane
d_B	Negative distance between the particle's centre of mass and a plane
D	Instantaneous oil slick diameter
D_0	Oil slick diameter at the initial state
$D_{SPH/RBC}$	Oil slick diameter provided by the SPH/ RBC simulation
D_{Fay}	Oil slick diameter predicted by Fay's equation
$\frac{D}{Dt}$	Lagrangian (or material) derivative
E	Reflection axis

g	Magnitude of gravity
h_o	Initial oil height
H	Ordinate of the fluid surface in the reservoir
k_1	Constant in Fay's equation
n	Unit vector normal to the plane
P_o	Point of the particle that will be in contact with the collision plane
P_I	Point of the particle in contact with the plane at the moment of collision
P	Absolute pressure
P_{mod}	Modified pressure
r	Particle radius
t	Time
t_o	Initial time instant
t_f	Final time instant of the gravity-inertial spreading
V	Volume of the oil spilled
v	Fluid velocity
$\mathbf{v}_{\mathrm{col}}$	Particle velocity immediately after the collision
$(\mathrm{v}_{\mathrm{col}})_N$	Magnitude of the component of the particle velocity perpendicular to the collision plane, immediately after the collision
$(\mathrm{v}_{\mathrm{col}})_T$	Magnitudes of the component of the particle velocity tangential to the collision plane, immediately after the collision
$\mathbf{v}_p$	Particle velocity before the collision
$(\mathbf{v}_p)_N$	Magnitude of the component of the particle velocity perpendicular to the collision plane before the collision
$\overrightarrow{X}$	Initial position
$\overrightarrow{X} + \Delta\overrightarrow{X}$	Final position
y	Ordinate of the fluid

Greek Symbols

Δt	Time step
Δ_{w}	Relation between oil and water densities
ρ	Fluid density
ρ_{w}	Water density
ρ_o	Oil density
υ	Fluid kinematic viscosity

Abbreviations

2D	Two-Dimensional
3D	Three-Dimensional
CDRA	Collision Detection and Response Algorithm
DPD	Dissipative Particle Dynamics
DSMC	Direct Simulation Monte Carlo
FEM	Finite Element Method
FPM	Finite Point Method
MD	Molecular Dynamics
MLS	Moving Least Square
MPS	Moving Particle Semi-Implicit
PFEM	Particle Finite Element Method
PI	Particle-in-Cell
RBC	Reflective Boundary Conditions
RKPM	Reproducing Kernel Particle Method
SPH	Smoothed Particle Hydrodynamics
SPH/RBC	Smoothed Particle Hydrodynamics Coupled with Reflective Boundary Conditions

List of Figures

List of Tables

Introduction

This chapter will present the conceptual differences between the molecular and continuum approaches and the current use of realistic physical Reflective Boundary Conditions (RBC) coupled with Lagrangian particle modelling to solve continuum mechanics problems.

1.1 Historical Background

Molecular modelling investigates molecular and atomic systems' structure, behaviour, and physical and mechanical properties. The interaction between atoms and molecules is taken into account. Quantum mechanics or classical methods can perform computer simulations at the molecular level. The latter can be deterministic, Molecular Dynamics (MD) or stochastic—like the Direct Simulation Monte Carlo (DSMC) [1–4].

In turn, continuum mechanics cannot represent the atoms and their bonds. The continuum hypothesis has validity (and is usually used) in models employed to solve engineering and physics problems in daily life [5]. Many molecules are associated with a small volume surrounding them in the continuum medium. A spatial point (that is also a material point) is defined for each one of those tiny volumes and has significant average physical quantities (density, temperature, viscosity, velocity, acceleration, internal energy, and others)—that are considered functions of position and time—and are governed by the classical laws of Physics. Due to the absence of voids in space, differential mathematics is used to write the classical physical laws. It is then possible to analyse the problems on a macroscopic scale without detailed knowledge of the molecular structure of matter [6, 7].

© The Author(s), under exclusive license to Springer Nature Switzerland AG 2025
C. A. D. Fraga Filho, *Reflective Boundary Conditions in SPH Fluid Dynamics Simulation*, Synthesis Lectures on Mechanical Engineering,
https://doi.org/10.1007/978-3-031-71582-2_1

Meshfree particle methods discretise the continuum medium in a set of particles (or nodal points) aiming to obtain a computational solution of the differential mathematics equations cited above. These methods originated over forty-five years ago. Among them, what appears to be the longest is the Smoothed Particle Hydrodynamics (SPH) Method developed by Lucy (1977) [8] and Gingold and Monaghan (1977) [9] to model astrophysical phenomena without boundaries, such as exploding stars and dust clouds. Other particle methods that can also be cited are Moving Particle Semi-Implicit (MPS), Moving Least Square (MLS), Reproducing Kernel Particle Method (RKPM), Finite Point Method (FPM), Particle-in-Cell (PI) and Particle Finite Element Method (PFEM) [10]. Currently, mesh-free particle methods are taught and applied in engineering, physics, mathematics, oceanography, environmental sciences and geosciences courses.

After performing the spatial discretisation of the domain by particles, the centre of mass of each Lagrangian element receives the coordinates that define its position in space, velocity, acceleration, density, internal energy, entropy, temperature and so on (these are the initial conditions). The mesh-free particle method will approximate the physical quantities at each time step (or numerical iteration) throughout the numerical simulation. Figure 1.1 shows the displacement of a particle considering two consecutive time steps of simulation that, in addition to their different spatial positions (visible), may still have other physical properties at each instant of time.

Whatever the boundary condition utilised, there is only one goal: to keep the Lagrangian particles of matter inside the domain, preventing them from crossing the contours. The concepts of molecular dynamics and continuum mechanics should be better understood in implementating Meshfree Lagrangian Particle Methods in problems of engineering, physics, biomechanics, and other areas at the macroscopic scale. The artificial boundary techniques or molecular dynamics models, still employed mainly in the

Fig. 1.1 The displacement of a Lagrangian particle in consecutive instants of time (t_o and $t_o + \Delta t$ are the initial and final times of simulation; $\vec{X}$ and $\vec{X} + \Delta \vec{X}$ are the initial and final positions of the particle)

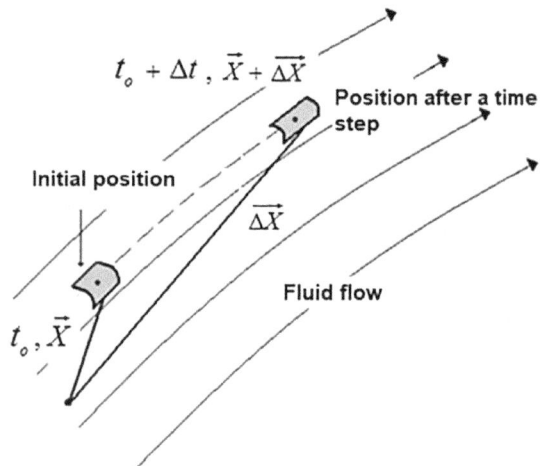

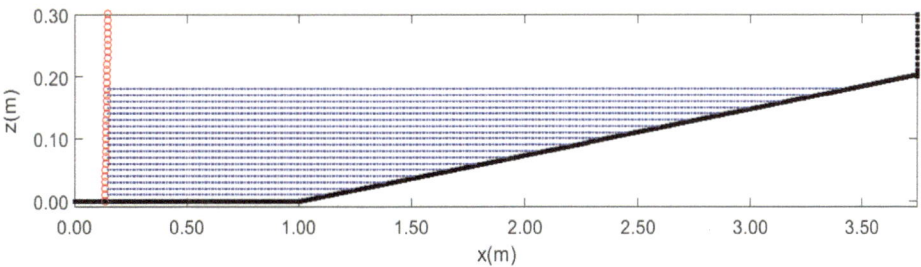

Fig. 1.2 The initial geometry and particle setup for the simulation of a propagation of a wave on a flat beach. Reproduced from [12], with the Springer Nature permission

macroscopic domain, despite leading to reasonable solutions for the problems analysed, generally introduce non-physical concepts for the evaluation of the interaction between fictitious particles and real fluid particles, or mixture concepts of molecular and continuum scales [6, 11].

For example, the author has recently studied the propagation of waves on a flat beach, using dynamic particles in the treatment of the contours [6, 12–14]. In this technique, the dynamic particles—disposed immediately after the domain—obey the physical laws of conservation that govern the behaviour of the real particles. They influence the physical properties and forces acting on the real particles and remain immobile throughout the simulation. In Fig. 1.2, the simulated geometry and the initial configuration of the particles is presented. The Lagrangian particles of water at the domain (in blue) and the boundary particles (black, representing the paddle, and red, representing the fixed bottom of the beach, and also delimiting the end of the domain) are seen. The simulation results agreed well with the analytical solution for the physical problem studied [12].

As said before, nowadays, the most used boundary conditions in particle simulation involve artificial computational techniques—in which ghost, dummy, repulsive and dynamic particles are placed on the boundary or in an immediate external region adjacent to the domain. In recent studies, there have been many efforts aiming to employ new techniques that do not use fictitious particles in the boundary treatment in the most advanced research groups, such as the semi-analytical model, with the evaluation of the contact forces in the rigid boundaries, the unified semi-analytical boundary conditions, an enhanced form of the previous technique with wall-corrected gradients, and possible to be used in a general boundary shape in two-dimensional domain (2D) and three-dimensional domain (3D); a boundary integral formulation for the contours treatment; and a new SPH modelling for viscous and non-viscous flows in the presence of 3D complex boundaries [6, 11, 13–15]. The replacement of these last boundary treatment techniques by RBC [6, 16, 17]—respecting the physical laws in the continuum domain, i.e. the continuum mechanics fundamentals—is a relevant advance in the search for a realistic treatment of the interaction between the fluid and the solid boundaries in mesh-free particle methods.

An appropriate boundary treatment became one of the most significant challenges in mesh-free particle methods. Due to this importance and the direct influence of the contours in obtaining consistent results for the physical properties of the particles near them, this book is focused on the presentation of the realistic physical reflective boundary conditions—based on the fundamentals of Physics and Analytical Geometry—their implementation and validation in the continuum domain, when coupled with the SPH Method—whose fundamentals can be found in [6, 18].

The implementation and validation of RBC coupled with the SPH method started in 2010 in the author's doctorate research. First, time has been devoted to developing and validating a 2D computational tool to simulate engineering problems. After the software testing and validation [19], the tool was used to simulate the problem of oil spreading oil on a calm sea. The forecast of the range of the pollutant in the first stage of the horizontal dispersion on a calm sea showed good agreement with the solution provided by the scientific literature [20, 21].

In 2018, the author (with the collaboration of colleagues) validated RBC for 3D domains and performed the first SPH simulations to study a water volume inside a still reservoir and dam breaking over a dry bed. The results were in agreement with the analytical solution (hydrostatic simulation) and with laboratory data provided by the literature (hydrodynamics study) [17].

Even more recently, results for a benchmark 3D fluid–structure interaction problem (FSI) were presented to the academic community [22, 23]. A water wave impacting a tall structure—a rigid obstacle fixed inside a reservoir—was simulated using the SPH method coupled with RBC. The results showed good agreement with the validated results of the literature. The applicability of RBC coupled with the SPH method for the fluid–structure interaction problem simulated encourages the implementation of the modelling to other FSI problems involving fixed and rigid boundaries.

1.2   Final Considerations

By extrapolating the continuum medium, there is a scientific effort to apply mesh-free particle methods coupled with other methods to bridge the gap between molecular and continuum approaches, which involve different scales in space and time. The mesoscopic scale is an intermediate scale between macroscopic and microscopic. The understanding and modelling of many phenomena require the development of studies in a challenging interface area, where the intersection of multiple disciplines, such as Physics, Chemistry, Material Science and Biology, occurs.

Hybrid molecular–continuum, multiscale and Dissipative Particle Dynamics (DPD) Models are recent computational methods that couple the events in the continuum and molecular/atomistic domains [11, 24–32].

We must distinguish the application of artificial boundary techniques in the continuum medium, as discussed in this book and previous literature [6, 11], with the studies dedicated to the bridge of molecular and continuum scales. On the contrary, the artificial boundary techniques, applied in the macroscopic scale, are generally used to solve problems from the empirical calibration of the parameters defined in their mathematical formulation.

1.3 Presentation of the Remaining Chapters

Chapter 2 presents the fundamentals of reflective boundary conditions, their implementation in the continuum domain, and the collision detection and response algorithm. Chapter 3 presents applications of RBC coupled with the SPH Method to solve problems in 2D domains. In turn, 3D simulations, the analysis of their results and discussions are shown in Chap. 4. Finally, Chapter 5 presents the conclusions about implementing RBC coupled with the SPH Method and perspectives on the future.

References

1. Rapaport, D.C. Molecular dynamics simulation: a tool for exploration and discovery using simple models, J. Phys.: Condens. Matter 26, 503104 (2014). https://doi.org/10.1088/0953-8984/26/50/503104
2. Alexander, F.J., Garcia, A.L. The Direct Simulation Monte Carlo Method. Computers in Physics 11, 588 (1997). https://doi.org/10.1063/1.168619.
3. Rapaport, D.C. The Art of Molecular Dynamics Simulation, 3rd edn. Cambridge University Press, UK (2004).
4. Alavi, S. Molecular Simulations: Fundamentals and Practice, 1st edn. Wiley, Germany (2020).
5. Batchelor, G.K. An Introduction to Fluid Dynamics, 3rd edn. Cambridge University Press, UK (2000).
6. Fraga Filho, C.A.D. Smoothed particle hydrodynamics fundamentals and basic applications in continuum mechanics. Springer Nature, Switzerland (2019).
7. Fox, R.W., McDonald, A.T., Pritchard, P.J. Introduction to Fluid Mechanics, 6th edn. Wiley, USA (2004).
8. Lucy, L.B. Numerical approach to testing the fission hypothesis. Astron. J. 82, 1013–1024 (1977). https://doi.org/10.1086/112164
9. Gingold, R.A., Monaghan, J.J. Smoothed particle hydrodynamics: theory and application to non-spherical stars. Mon. Not. R. Astron. Soc. 181, 375–389 (1977). https://doi.org/10.1093/mnras/181.3.375
10. Fraga Filho, C.A.D., Schuina, L.L., Porto, B.S. An Investigation into Neighbouring Search Techniques in Meshfree Particle Methods: An Evaluation of the Neighbour Lists and the Direct Search. Arch Computat Methods Eng 27, 1093–1107 (2020). https://doi.org/10.1007/s11831-019-09345-9

11. Fraga Filho, C.A.D. On the boundary conditions in Lagrangian particle methods and the physical foundations of continuum mechanics. Continuum Mech. Thermodyn. 31, 475–489 (2019). https://doi.org/10.1007/s00161-018-0702-2
12. Filho, C.A.D.F., Piccoli, F.P. Diffusive terms applied in smoothed particle hydrodynamics simulations of incompressible and isothermal Newtonian fluid flows. J Braz. Soc. Mech. Sci. Eng. 43, 479 (2021). https://doi.org/10.1007/s40430-021-03158-3
13. Fraga Filho, C.A.D., Chacaltana, J.T.A. Boundary Treatment Techniques in Smoothed Particle Hydrodynamics: Implementations in Fluid and Thermal Sciences and Results Analysis. Interdisciplinary Journal of Engineering Research—RIPE. In: Proceedings of the XXXVII Iberian Latin American Congress on Computational Methods in Engineering—CILAMCE 2016, Brasília-DF (2017). https://periodicos.unb.br/index.php/ripe/article/view/21270
14. Crespo, A.J.C., Gómez-Gesteira, M., Dalrymple, R.A. Boundary conditions generated by dynamic particles in SPH methods. CMC Comput. Mat. Cont. 5(3), 173–184 (2007). https://doi.org/10.3970/cmc.2007.005.173
15. Rezavand M., Zhang C., Hu X. Generalized and efficient wall boundary condition treatment in GPU-accelerated smoothed particle hydrodynamics. Comput Phys Commun 281, 108507 (2022). https://doi.org/10.1016/j.cpc.2022.108507
16. Fraga Filho, C.A.D. An algorithmic implementation of physical reflective boundary conditions in particle methods: collision detection and response. Phys. Fluids 29, 113602 (2017). https://doi.org/10.1063/1.4997054
17. Fraga Filho, C.A.D., Peng, C., Islam M.R.I., McCabe, C., Baig, S., Durga Prasad G. V. Implementation of three-dimensional physical reflective boundary conditions in mesh-free particle methods for continuum fluid dynamics: Validation tests and case studies. Physics of Fluids 31:103606 (2019). https://doi.org/10.1063/1.5115776
18. Liu, G.R., Liu, M.B. Smoothed Particle Hydrodynamics: a Meshfree Particle Method. World Scientific, Singapore (2003).
19. Fraga Filho, C.A.D. Development of a computational instrument using a lagrangian particle method for physics teaching in the areas of fluid dynamics and transport phenomena. Rev Bras Ensino Fís 39(4):e4401 (2017). https://doi.org/10.1590/1806-9126-rbef-2016-0289
20. Fraga Filho, C.A.D. A SPH Model for Prediction of Oil Slick Diameter in the Gravity-inertial Spreading Phase. In: Proceedings of the V International Conference on Particle-based Methods – Fundamentals and Applications-PARTICLES 2017, Hannover, Germany (2017). Available at https://core.ac.uk/reader/323499034, accessed on 31 January 2023.
21. Fraga Filho, C.A.D. A Lagrangian analysis of the gravity-inertial oil spreading on the calm sea using the reflective oil-water interface treatment. Environ Sci Pollut Res 28, 17170–17180 (2021). https://doi.org/10.1007/s11356-020-11508-2
22. Fraga Filho, C. A. D. Fluid-structure interaction simulation by SPH and reflective boundary conditions, in B.H.V. Topping, J. Kruis, (Editors), Proceedings of the Fourteenth International Conference on Computational Structures Technology, Civil-Comp Press, Edinburgh, UK, Online volume: CCC 3, Paper 3.1. Available at https://www.ctresources.info/ccc/paper.html?id=9393. Accessed on 15 September, 2023.
23. Fraga Filho, C.A.D. Reflective boundary conditions coupled with the SPH method for the three-dimensional simulation of fluid–structure interaction with solid boundaries. J Braz. Soc. Mech. Sci. Eng. 46, 256 (2024). https://doi.org/10.1007/s40430-024-04807-z
24. Li, L., Shen, L., Nguyen, G.D. et al. A smoothed particle hydrodynamics framework for modelling multiphase interactions at meso-scale. Comput Mech 62, 1071–1085 (2018). https://doi.org/10.1007/s00466-018-1551-3
25. Fish, J., Wagner, G.J., Keten, S. Mesoscopic and multiscale modelling in materials. Nat. Mater. 20, 774–786 (2021). https://doi.org/10.1038/s41563-020-00913-0

26. Coveney, P.V., Boon, J.P., Succi, S. Bridging the gaps at the physics–chemistry–biology interface. Philos. Trans. R. Soc. A 3, 1–2 (2016). https://doi.org/10.1098/rsta.2016.0335
27. Delgado-Buscalioni, R., Coveney, P.V., Riley, G.D., Ford, R.W. Hybrid molecular-continuum fluid models: implementation within a general coupling framework. Philos. Trans. R. Soc. A 363(1833), 1975–85 (2005). https://doi.org/10.1098/rsta.2005.1623
28. Mukhopadhyay, S., Abraham, J. A particle-based multiscale model for submicron fluid flows. Phys. Fluids 21, 027102 (2009). https://doi.org/10.1063/1.3073041
29. Stalter, S., Yelash, L., Emamy, N., Statt, A., Hanke, M., Lukáčová-Medvid'ová, M., Virnau, P. Molecular dynamics simulations in hybrid particle-continuum schemes: pitfalls and caveats. Comput. Phys. Commun. 224, 198–208 (2018). https://doi.org/10.1016/j.cpc.2017.10.016
30. Sih, G.C. (ed.) Multiscaling in Molecular and Continuum Mechanics: Interaction of Time and Size from Macro to Nano: Application to Biology, Physics, Material Science, Mechanics, Structural and Processing Engineering. Springer, Dordrecht (2007). https://doi.org/10.1007/978-1-4020-5062-6
31. Petsev, N.D., Leal, L.G., Shell, M.S. Multiscale simulation of ideal mixtures using smoothed dissipative particle dynamics. J. Chem. Phys. 144, 084115 (2016). https://doi.org/10.1063/1.4942499
32. Liu, M.B., Liu, G.R., Zhou, L.W., Chang, J.Z. Dissipative particle dynamics (DPD): an overview and recent developments. Arch. Comput. Methods Eng. 22(4), 529–556 (2015). https://doi.org/10.1007/s11831-014-9124-x

Reflective Boundary Conditions

2

This chapter will present the fundamentals of the reflective boundary conditions and the collision detection and response algorithm (the computational tool developed to implement the boundary technique).

2.1   Fundamentals

Nowadays, significant scientific effort is focused on bridging the gap between molecular and continuum approaches, involving different scales in space and time.

This book presents reflective boundary conditions coupled with a Lagrangian particle method. The particle method used in the case studies discussed here was Smoothed Particle Hydrodynamics (SPH), whose foundations and other aspects are available, for example, in the literature references [1, 2]. The central attention of this work is focused on the presentation of the applications of reflective boundary conditions in problems in the two- and three-dimensional domains, whose deeper understanding can be carried out by the reader in the references cited.

When a fluid is flowing, there are two primary sources of energy loss. The first is due to friction between the fluid's layers (viscosity effect), and the second is due to the fluid's contact with the walls of the reservoir, channel, or pipes [3]. The coefficient of friction (CF) ranges from 0.0 to 1.0 and represents the percentage of tangential velocity lost during a contact. A higher CF value results in a more significant loss of tangential velocity, suggesting a rougher surface. When CF equals 1.0, the particle's tangential velocity is entirely gone. Elasticity is connected with the energy returned to the particle following the impact. The coefficient of restitution of kinetic energy (CR) determines the fraction of

© The Author(s), under exclusive license to Springer Nature Switzerland AG 2025   9
C. A. D. Fraga Filho, *Reflective Boundary Conditions in SPH Fluid Dynamics Simulation*, Synthesis Lectures on Mechanical Engineering,
https://doi.org/10.1007/978-3-031-71582-2_2

kinetic energy recovered. CR values near zero suggest highly inelastic collisions, whereas a CR value of one indicates a perfectly elastic collision. Elasticity influences particle velocity in the direction perpendicular to the contact site [4]. Particles are smooth, rigid, and spherical with non-zero masses characterised by their volumes and fluid density. Newton's Laws are applied to physically treat the particle collisions against the boundaries in the continuum domain [4–6]. The rest of this Chapter will present realistic physical reflective boundary conditions (RBC) for treating problems on the macroscopic scale.

To obtain the input data for the Collision Detection and Response Algorithm (CDRA), the first step is to solve the momentum equation (Newton's second law of motion) using the appropriate Lagrangian particle method, which provides the acceleration of the particles at each time step. Then, the final positions of the particles' centre of mass and velocities are obtained using a temporal integration method [6, 7]. If the integration time step is sufficiently small, the particle's acceleration and velocity can be considered constant at each numerical iteration. The simplest integration method (first-order Runge–Kutta or Euler's Method) provides, at each time step, the particle's position from its current derivative (velocity) multiplied by the time interval and the particle's velocity for the next iteration by its current derivative (acceleration calculated using the momentum equation) multiplied by the time step.

Figure 2.1 depicts the particle's displacement according to the Euler's method and the more precise particle's displacement promoted by a higher-order integration method in a time step (Δt). If the time step is too large, the differences between the positions (the actual position and that predicted by Euler) may be considerable. In this book, the Euler method was used with good accuracy in small simulation time steps, where the particle was thought to move linearly along the velocity direction.

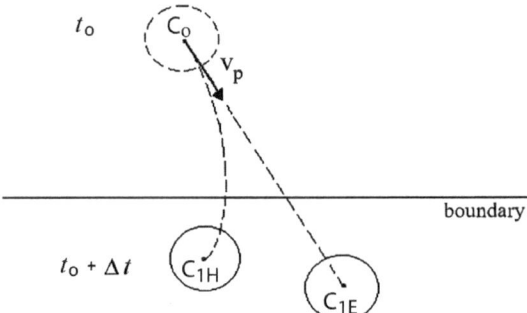

Fig. 2.1 Numerical integration using the Euler method (E) and a higher-order method (H). The particle's position and velocity at the initial time instant are C_0 and $\mathbf{v}_p$, and the positions of the particle's centres of mass at the end of the numerical iteration are given by C_{1E} and C_{1H}, respectively. The particle's movement was done freely, not considering the boundary's existence. Reproduced from [6], with the permission of AIP Publishing

2.2   Collision Detection and Response Algorithm

This section will explain how the Collision Detection and Response Algorithm (CDRA) is implemented. The algorithm requires the following input data:

1. The equations of the planes that define the geometry of the problem. These equations are obtained using principles of analytical geometry. The normal vectors of these planes must be normalised.
2. The initial and final positions of the centres of mass of all particles on each numerical iteration. At the initial time, each particle must be located in the positive region of the domain.
3. The particles' velocities at each time step.

The physical quantities of items 2 and 3 are obtained by performing temporal integration operations.

2.2.1   Particles' Collisions Detection

At each numerical iteration, the following steps will be performed to detect the particles' collisions against the boundaries:

1. Calculate the distance from the particle to the plane. The plane has a positive and a negative region, with the positive side being the direction in which the normal unit vector $\mathbf{n}$ points. Initially, the particle will always be in the positive region of the plane.

   In Fig. 2.2, point A is on the side of the plane towards which the normal unit vector $\mathbf{n}$ points; therefore, d_A is a positive and d_B is a negative distance. This is the side of the plane where the particle collides. On the other hand, point B is on the negative side of the plane.
2. Collision detection: there are three possible particle positions about the plane in the case of the free particle's movement. They are:

Fig. 2.2 Positive and negative sides of the plane (where the centres of mass A and B are located, respectively). Reproduced from [6], with the permission of AIP Publishing

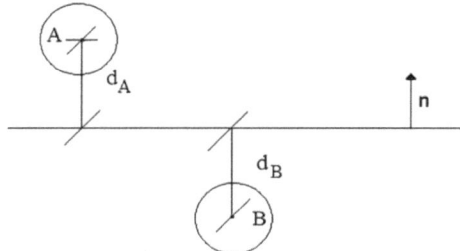

1. The particle has not reached the spatial position of the plane.
2. The particle has reached the spatial position of the plane, but did not completely surpass it.
3. It has passed through the plane entirely.

Figure 2.3 illustrates three potential scenarios. In this diagram, "d" represents the distance between the particle's centre of mass and the plane, "r" is the particle's radius, t_o indicates the initial time, and "Co" and "C_1" denote the positions of the centres of mass at the beginning and end of the numerical iteration (at the time $t_o + \Delta t$) achieved through numerical integration for free particle motion in the absence of obstacles.

Considering its position in space, it should now be checked whether the particle will collide against the plane at the end of its trajectory. A collision can occur in two situations: when d > 0 and d < r, or for d < 0. In Fig. 2.3 (A), d < 0 and d > r; a collision with the plane has not occurred. In (B), d is positive and smaller than the radius, and in (C) d is negative and the particle is on the negative side of the plane. In both situations, a collision has occurred.

The simulated domain is normally composed of some rigid boundaries (as will be seen in the study cases of this book). Thus, the determination of the impact plane—which has the point of contact with the particle ("P_1") closest to "P_o"— is done from the analysis of the particle trajectory. All the possible collision planes are stored if the distance between "C_1" and a plane indicates a possible collision, as shown in Fig. 2.3b, c. The length of the line segment P_1–P_o is calculated for all the possible particle impact planes. The plane for

Fig. 2.3 Collision test for a particle in relation to the lower plane (calculated from the analysis of the particle trajectory over a single time step). Reproduced from [6], with the permission of AIP Publishing

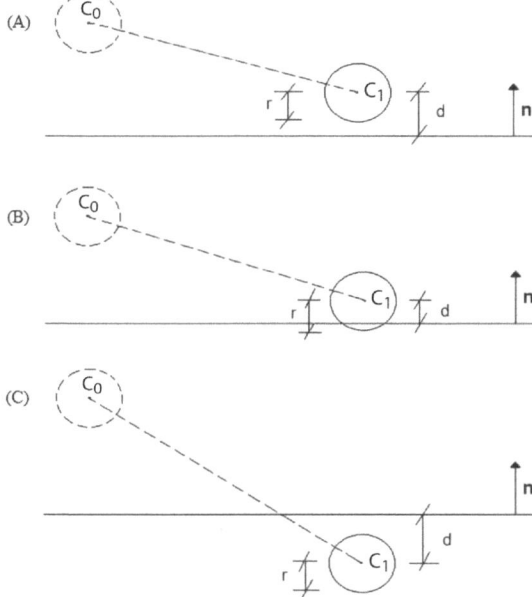

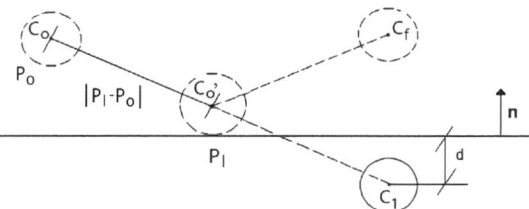

Fig. 2.4 The smallest value of the length of the line segment P_I–P_O defines the collision plane. The diagram shows this plane and the position of the centre of mass of the particle after the collision response. Reproduced from [6], with the permission of AIP Publishing

which the length of the line segment is the smallest will be the collision plane, as shown in Fig. 2.4.

2.2.2   Particles' Collisions Response

If a collision against the boundary has occurred, the particle must be relocated within the domain (enforcement of the non-penetration condition). The CDRA algorithm performs the particle reflection considering the detected impact plane. After reflection, the particle trajectory becomes different from that at the beginning of the process. The initial point of the new trajectory is the centre of mass at the moment of contact with the plane of collision (C_o') and the endpoint is point C_f, as shown in Fig. 2.4.

Figure 2.5 shows the parameters involved in the particle reflection, namely: r is the particle radius, d is the distance between the particle's centre of mass and the plane (negative when the centre of mass crosses the plane), "C_o" is the initial position of the centre of mass of the particle, "C_1" is the position of the centre of mass, in a motion without obstacles, at the end of the numerical iteration, "C_f" is the position of the centre of mass obtained after the collision response, **n** is the unit vector normal to the collision plane and E is the reflection axis.

After detecting the particle collision with the boundary, the velocity and position of the particle's centre of mass must also be corrected. The coordinates of the particle's centre of mass perpendicular to the collision plane must be corrected when a reflection occurs, as shown in Eq. (2.1).

$$(C_f)_N = (C_1)_N + (1.00 + CR)(r - d) \tag{2.1}$$

Figure 2.6 depicts the corrected velocities of the particle immediately after the collision.

The magnitude of the velocity component perpendicular to the collision plane, immediately after the impact is given by Eq. (2.2):

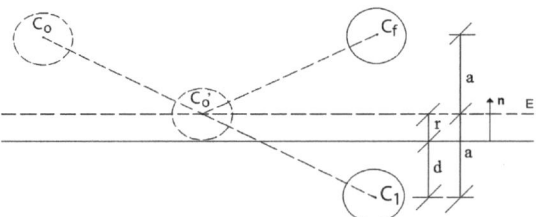

Fig. 2.5 The reflection axis is given by E. The distance between "C_1" and the axis E is equal to the distance between this axis and "C_f" for a totally elastic collision. Reproduced from [6], with the permission of AIP Publishing

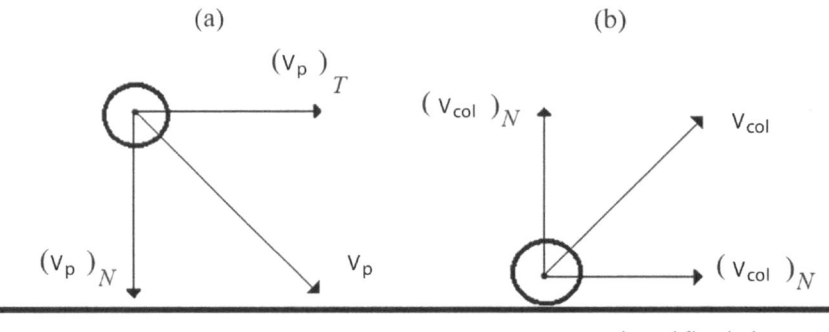

rough and fixed plane

Fig. 2.6 The velocities of an oil particle before the impact against the rough and rigid plane fixed on the water surface (**a**) and immediately after it (**b**). Reproduced from [4], with the permission of AIP Publishing

$$(\mathrm{V_{col}}) = \mathrm{CR} \times \left(\mathrm{V_p}\right)_N \tag{2.2}$$

In the direction tangential to the collision plane we have, Eq. (2.3):

$$(\mathrm{V_{col}})_T = (1 - \mathrm{CF}) \times \left(\mathrm{V_p}\right)_T \tag{2.3}$$

The direction of the normal velocity component changes after the collision, while the tangential velocity component stays the same in every scenario.

Once the points defining the new particle trajectory are known, i.e. "C_o'" and " C_f ", the steps presented in the last paragraph of Sect. 2.2.1 and in this subsection must be repeated until no more collisions occur against a plane (the particle's centre of mass is repositioned within the domain in the current iteration).

The steps described in Sect. 2.2 must be carried out for each particle in the domain.

2.2.3 Computer Code

Figure 2.7 presents a computer code used in the SPH method simulation coupled with the reflective boundary conditions.

The routines used in each simulation depend on the case studied. The literature [1, 4, 5, 8] provides more detailed information on the numerical simulations. Below is a brief presentation of the computer code and its routines.

1. Geometry/Computational domain definition.
2. Initial conditions definition: The initial positions, velocities, densities, temperatures, support radius and other fluid particles' physical properties are set at the beginning of the simulation.
3. Kernel definition: The smoothing function used in interpolations is chosen.
4. Pressure calculus: This routine updates the pressure field acting on the particles.
5. Searching for the neighbouring particles: The particles within the domain of influence of a reference particle may vary over time, and the search must be performed at each numerical iteration.
6. Mass conservation equation solution: The particles' density calculus is conducted for each particle at the domain.
7. Density renormalisation: Numerical correction commonly made in SPH simulations.
8. Surface forces calculation: approximations of surface forces (pressure and viscous force) acting on the particles are obtained.
9. Pressure gradient corrections: Numerical corrections commonly made in SPH simulations.
10. Application of external forces: External forces, such as gravity and free surface forces, are applied to the particles.
11. Energy conservation equation solution.
12. Momentum balance equation solution: Particles' acceleration field is obtained.
13. Temporal integration: Predicting the particles' quantities to the next iteration. The Courant-Friedrichs-Lewy (CFL) stability criterion [9] is applied in the time-step setting to ensure convergence of results.
14. Application of the reflective boundary conditions (CDRA).
15. Output files: These files are obtained at the end of each numerical iteration, and from their data, graphical representations of the fluid's physical properties are generated.
16. Accuracy: This routine verifies whether the desirable accuracy has been achieved or if a new iteration must be executed.

Fig. 2.7 A generic flowchart
showing the routines used in
the SPH/ RBC simulations.
The active routines are defined
according to the problem
studied

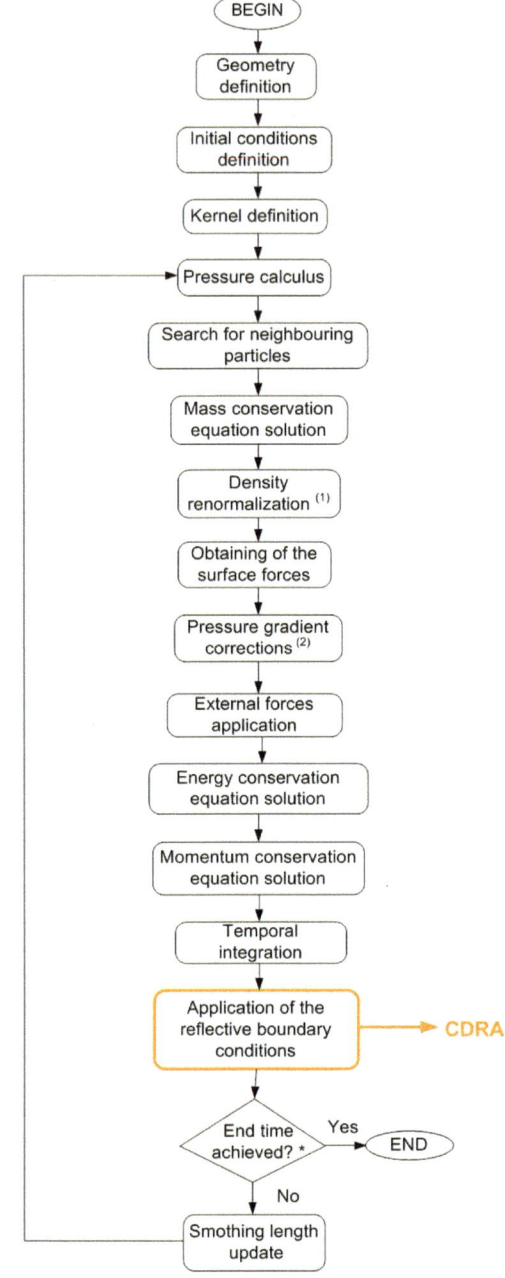

(1) and (2) are numerical corrections usually used in the SPH simulation [1].

2.2.4   Final Considerations

2.2.4.1   CDRA

The collision detection and response algorithm (CDRA) detects the particles' collisions against the boundaries and performs the reflections into the domain (no-penetration condition). Once the geometry of the domain, the particle velocity, and the initial and final position of the centre of mass (at the beginning and the end of the numerical iteration, respectively) are known, these data will be used as input data in CDRA. The algorithm considers the interactions between particles and boundaries to be minimal and negligible instants of time. The algorithm step-by-step is presented below.

CDRA Step-By-Step

1. **for** each particle in the domain **do**
   2. Calculate the distances between the centres of mass of the particles (end of the trajectory at every numerical iteration) and each plane
   3. Check for possible collision planes
   4. Identify the plane of collision by the criterion of the smallest trajectory of the particle
   5. Reflect the centre of mass of the particle considering the collision plane identified
   6. Perform the correction of the particle's velocity
   7. Verify that the particle returned to the interior of the domain after reflection
   8. **if** particle did not return to the domain **then**
      9. Repeat steps 2 through 7 until the particle return to the domain (in case of collisions against more than one plane in the time step analysed)

As output data, we have the positions of the centre of mass and corrected particles' velocities at the final instant of each numerical iteration.

The literature presents the implementation of the RBC, for two- and three-dimensional domains [4, 6]. In Appendixes there are RBC validations tests for two- and three-dimensional domains.

2.2.4.2   Collisions Between Fluid Particles

In the fluid dynamics field, the simulation of interparticle collisions inside the fluid is generally not included in the Lagrangian particle simulation due to the high computational cost [10]. The one-particle model—in which each mass element (or Lagrangian particle) moves through space without regard to the movement of the other—is often used for efficient computation and still provides good and acceptable results, as was seen in the problems addressed in this book.

On the other hand, in solid mechanics, it is crucial to consider the interactions between particles' surfaces, and contact algorithms need to be implemented [11, 12].

References

1. Fraga Filho, C.A.D. Smoothed particle hydrodynamics fundamentals and basic applications in continuum mechanics. Springer Nature, Switzerland (2019).
2. Liu G. R., Liu M. B. Smoothed Particle Hydrodynamics: a Meshfree Particle Method. World Scientific, Singapore, 2003.
3. Hanot S., Belushkin M., Foffi G.. Partial slip at fluid–solid boundaries by multiparticle collision dynamics simulations. Soft Matter, 9, 291–296 (2013). https://doi.org/ https://doi.org/10.1039/c2sm26316e
4. Fraga Filho, C.A.D., Peng C., Islam R.I., McCabe C., Baig B., Venkata Durga Prasad, G.V.D. Implementation of three-dimensional physical reflective boundary conditions in mesh-free particle methods for continuum fluid dynamics: Validation tests and case studies. Phys. Fluids 31, 103606 (2019). https://doi.org/10.1063/1.5115776
5. Fraga Filho, C.A.D. Reflective boundary conditions coupled with the SPH method for the three-dimensional simulation of fluid–structure interaction with solid boundaries. J Braz. Soc. Mech. Sci. Eng. 46, 256 (2024). https://doi.org/10.1007/s40430-024-04807-z
6. Fraga Filho, C.A.D. An algorithmic implementation of physical reflective boundary conditions in particle methods: Collision detection and response. Physics of Fluids 29, 113602 (2017). https://doi.org/10.1063/1.4997054
7. House, D. H., Keyser, J.C. Foundations of Physically Based Modeling and Animation. CRC Press, Taylor & Francis Group, Boca Raton, Florida, USA (2017)
8. Fraga Filho, C. A. D. Development of a computational instrument using a Lagrangian particle method for physics teaching in the areas of fluid dynamics and transport phenomena. Rev. Bras. Ensino Fís. 39 (4) (2017). https://doi.org/10.1590/1806-9126-RBEF-2016-0289
9. Courant R., Friedrichs K., Lewy H. On the partial difference equations of mathematical physics. IBM Journal 11, 215–234 (1967).
10. Korzilius S. P., Kruisbrink A. C. H., Schilders W. H. A., Anthonissen M. J. H., Yue T. Momentum conserving methods that reduce particle clustering in SPH. (CASA-report; Vol. 1415). Technische Universiteit Eindhoven (2014) Available at https://pure.tue.nl/ws/files/3858217/376670351851652.pdf . Accessed on July 09, 2024.
11. Campbell J., Vignjevic R., Libersky L. A contact algorithm for smoothed particle hydrodynamics. Comput. Methods Appl. Mech. Eng. (2000). https://doi.org/10.1016/S0045-7825(99)00442-9
12. Seo S., Min O., Lee J. Application of an improved contact algorithm for penetration analysis in SPH. Int. J. Impact Eng. (2008). https://doi.org/10.1016/j.ijimpeng.2007.04.009

Applications of Reflective Boundary Conditions in Two-Dimensional Domains

3

The first applications of RBC (in two-dimensional domains) in hydrostatics and hydro-dynamics, as well as their validations and discussion of results, will be presented in this chapter.

3.1 Uniform and Incompressible Fluid at Rest Inside a Reservoir

This hydrostatic problem involves a stationary tank filled with liquid that is open to the atmosphere. The transverse section of the tank measures 1.0 m × 1.0 m. The 2,500 water inside the reservoir was at 20 °C and sea level ($\rho = 1.00 \times 10^3$ kg/m^3, ν_{water} $\nu = 1.00 \times 10^{-6}$ m^2/s) and was considered isothermal. The frame of reference originates at the bottom of the tank, with a positive vertical orientation pointing upwards. Figure 3.1 shows the distribution of 50 Lagrangian discretisation particles on each tank side (the black points represent their centre of masses).

In hydrostatic problems, ensuring the balance of forces and the consistent positions of particles over time is challenging when using the standard SPH formulation in conjunction with the virtual particles boundary technique [1–3]. This difficulty arises from the errors involved in approximating the physical properties using the SPH Lagrangian method [4–7].

Using the modified pressure concept [8], the balance of forces was guaranteed as shown in Eq. (3.1):

$$P_{\text{mod}} = (P - P_{\text{o}}) - \rho g (H - y) \tag{3.1}$$

© The Author(s), under exclusive license to Springer Nature Switzerland AG 2025
C. A. D. Fraga Filho, *Reflective Boundary Conditions in SPH Fluid Dynamics Simulation*, Synthesis Lectures on Mechanical Engineering,
https://doi.org/10.1007/978-3-031-71582-2_3

Fig. 3.1 Initial particle setup inside the reservoir. Reproduced from [4], with the permission of AIP Publishing

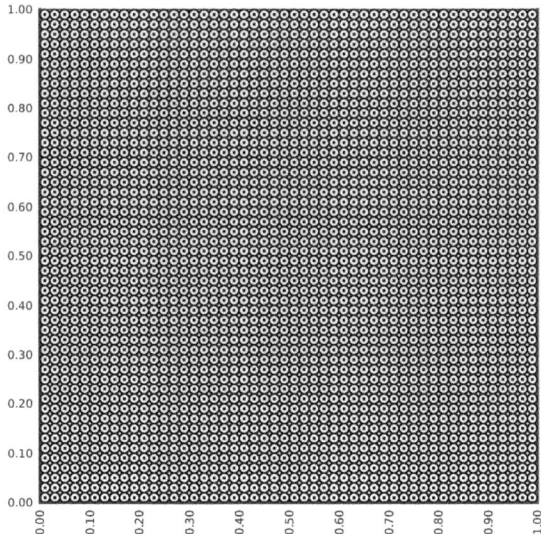

where:

P_{mod} is the modified pressure

P is the absolute pressure

P_{o} is the reference pressure

ρ is the fluid density

g is the magnitude of gravity

H is the ordinate of the fluid surface in the reservoir

y is the ordinate of the fluid (vertical direction).

Considering that the pressure exerted by the fluid column on the particle with ordinate y, the modified pressure is zero on each particle in the domain. Using the modified pressure the form of the momentum balance equation is shown in Eq. (3.2).

$$\frac{\mathrm{D}\mathbf{v}}{\mathrm{Dt}} = -\frac{\nabla P_{\mathrm{mod}}}{\rho} + \upsilon\nabla^2\mathbf{v} \qquad (3.2)$$

After solving the conservation of mass and momentum equations and updating the positions of the centres of mass and velocities of the fluid particles using time integration at each numerical iteration, we found that these physical properties remained unchanged, and the system maintained the hydrostatic equilibrium.

When using the CDRA, the contact between fluid particles and the lateral and lower planes was detected (that is, the distance between the centres of mass of those particles and the plane was equal to the particle radius). During the response stage, particles' reflections were carried out based on Eq. (2.9) with no change in the positions of the centres of mass. A zero coefficient of restitution of kinetic energy was used, and the particle velocities' responses were also zero. The simulation results obtained using the collision detection and response algorithm aligned perfectly with Newton's physical laws.

3.2   Dam-Breaking Over a Dry Bed

Studies on dam-breaking are crucial in preventing accidents, as they can result in severe environmental consequences and threaten residents near those facilities. In this problem, the fluid is assumed to be incompressible, uniform, and isothermal. The solution of the Navier–Stokes equations, mass and momentum conservation, is necessary to obtain the particles' acceleration and, shortly after, the positions and velocities of the particles (input data for the algorithm) after performing the temporal integration. Figure 3.2 depicts the simulated computational domain and the initial particle setup.

The reservoir had dimensions of 0.42 m in length and 0.44 m in height. Numerical simulations were conducted for a rectangular section of water area with a height of 0.228 m and a width of 0.114 m. The simulations used 2,556 fluid particles with a distance of 3.26 mm between their centres of mass. The fluid's density and kinematic viscosity were $\rho = 1.00 \times 10^3$ kg/m^3 and $\upsilon = 1.00 \times 10^{-6}$ m^2/s, respectively. The fluid's absolute pressure was calculated as the sum of a hydrostatic and a dynamic component, as predicted by Tait's state equation [9]. The free surface particles were marked, and the constant pressure condition (zero) was applied and renewed at every numerical iteration (Newman boundary conditions) [10]. The time step used was 1.00×10^{-5} s. The simulations were run for a simulated time of 0.30 s (30,000 numerical iterations) and were validated using

Fig. 3.2 The simulated two-dimensional computational domain and the initial particle setup. The fluid particles' centres are shown in the figure

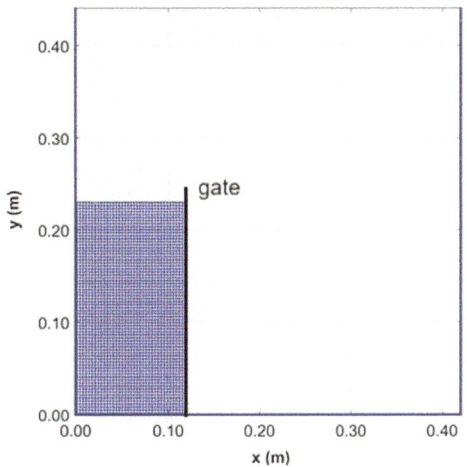

Fig. 3.3 Experimental and numerical results (in the second line) for the dam-breaking simulation. Reproduced from [12], with the permission of AIP Publishing

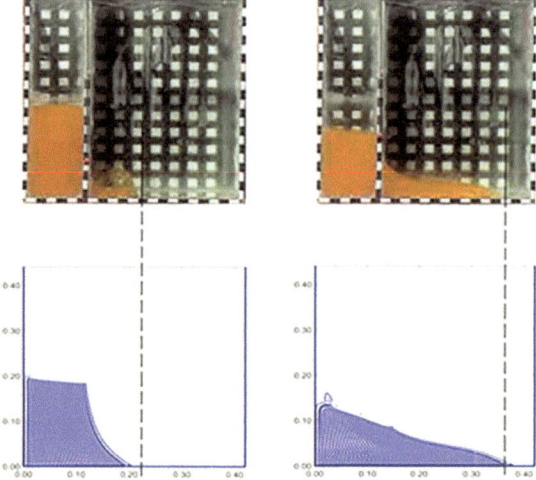

experimental data [11]. More information on the SPH simulation performed can be found in [12].

In the numerical simulation, the dam gate was suddenly opened. The water flow started immediately after that initial time instant. After determining the particle accelerations using the momentum conservation equation and performing temporal integration using Euler's method, the positions of the particles' centres of mass (input data for the algorithm) were obtained at the end of each numerical iteration. The energy restitution coefficient was 1.00, and there was no friction coefficient in the operations carried out by CDRA. The wavefront position was monitored before its collision with the right-hand reservoir wall.

The largest disparities between the positions of the wavefront (as measured experimentally and provided by the SPH Lagrangian Method using reflective boundary conditions) were observed at the beginning of the dam-breaking. At 0.10 s, the difference was 6.90% (0.2032 m numerical/0.2170 m experimental). As time passed, the difference decreased, and at 0.20 s, it was 4.40% (0.3776 m numerical/0.3610 m experimental). In this way, numerical results agreed well with the experimental ones in the simulated period. Figure 3.3 provides a visual representation of the experimental and numerical results.

3.3 Oil Spreading on a Calm Sea

The spread of an oil slick occurs because the pollutant tends to flow over itself. This is the most critical transport process in the first few hours of a spill. Understanding and quantifying this process is crucial to protecting the environment. Knowledge of physical quantities related to pollutant transport, such as velocities and positions, is essential for taking timely measures to safeguard the environment.

The numerical simulation results contribute to developing contingency and emergency plans, which are required by law. Predicting the reach of the oil slick over time allows quick and effective measures to be taken to protect coastal areas.

Fay [13, 14] fitted curves to experimental oil spreading data, considering a calm sea condition, that is, without winds, tidal currents and waves. In the first phase of spreading (gravity-inertial), gravity predominates as the driving force, and inertia is the main force resisting spreading.

The SPH/ RBC numerical results obtained in this work's simulation were validated from Fay's equation for gravity-inertial spreading. According to this researcher, on a calm condition, the oil spreads in a circular shape, in a displacement that presented radial symmetry, according to Eqs. (3.3):

$$D = 2k_1\left(\Delta_w g V t^2\right)^{\frac{1}{4}} + D_o,$$

$$\Delta_w = \frac{\rho_w - \rho_o}{\rho_w} \tag{3.3}$$

where:

D_o is the oil slick diameter at the initial state

D is the oil slick diameter in the instant t

ρ_w is the water density

ρ_o is the oil density

g is the magnitude of the gravitational acceleration

V is the volume of the oil spilled

$k_1 = 1.14$

This stage of the spreading lasts until the time instant t_f given by Eq. (3.4):

$$t_f = \left(\frac{h_o}{g\Delta_w}\right)^{\frac{1}{2}} \tag{3.4}$$

where h_o is the initial oil height.

According to Stolzenbach [15], the initial movement of the spilled oil is similar to that of a dam break. This study considered this hypothesis, and the Smoothed Particle Hydrodynamics (SPH) method was used to solve the oil phase's mass and momentum conservation equations [16].

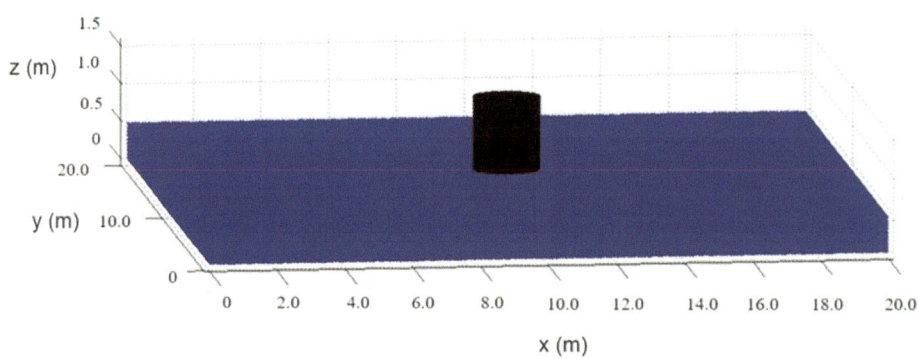

Fig. 3.4 Simulated geometry showing the oil cylinder over the water surface at the initial time instant

The simulated domain was a tank 20 m in length, 20 m in width, and 1.5 m in height. The water level was 0.50 m, and initially, an oil volume disposed of in the centre of the tank was a cylinder with a height and diameter of 1 m. Figure 3.4 depicts the geometry simulated and the oil cylinder over the water surface, which was subjected to horizontal dispersion (spreading).

The initial lateral distances between the particles' mass centres were 2.00×10^{-2} m. Two thousand five hundred oil particles and 26,000 water particles (density of 1,000 kg/m^3 and absolute viscosity of 1.00×10^{-3} Pa.s) were employed in the discretisation of the domain. The oil was considered a homogeneous, uniform and isotropic fluid. The simulations were carried out for light-weight crude oil, with a density of 850.00 kg/m^3 and an absolute viscosity of 3.32×10^{-3} Pa.s. The duration of the oil spreading in its first stage, obtained by Eq. (3.4), was 0.82 s with a total of 5,815 numerical iterations.

In the modelling used, the interaction between the two phases (water and oil) was measured from the definition of a horizontal rough and rigid plane fixed at over the water level as shown in Fig. 3.5, over which the oil spreads.

In the simulations, the movement of oil particles on the water's surface was monitored throughout. In each numerical iteration, the spreading range was defined by a circumference with a radius equal to the farthest distance a particle (represented by its centre of mass) reached, measured from the centre of the tank (10.00 m, 10.00 m, 0.50 m). Figure 3.5a displays the radius that defines the spreading range.

Figure 3.6 schematically shows the temporal evolution of the circular-shaped oil slick on a calm sea condition (from an upper view of the reservoir).

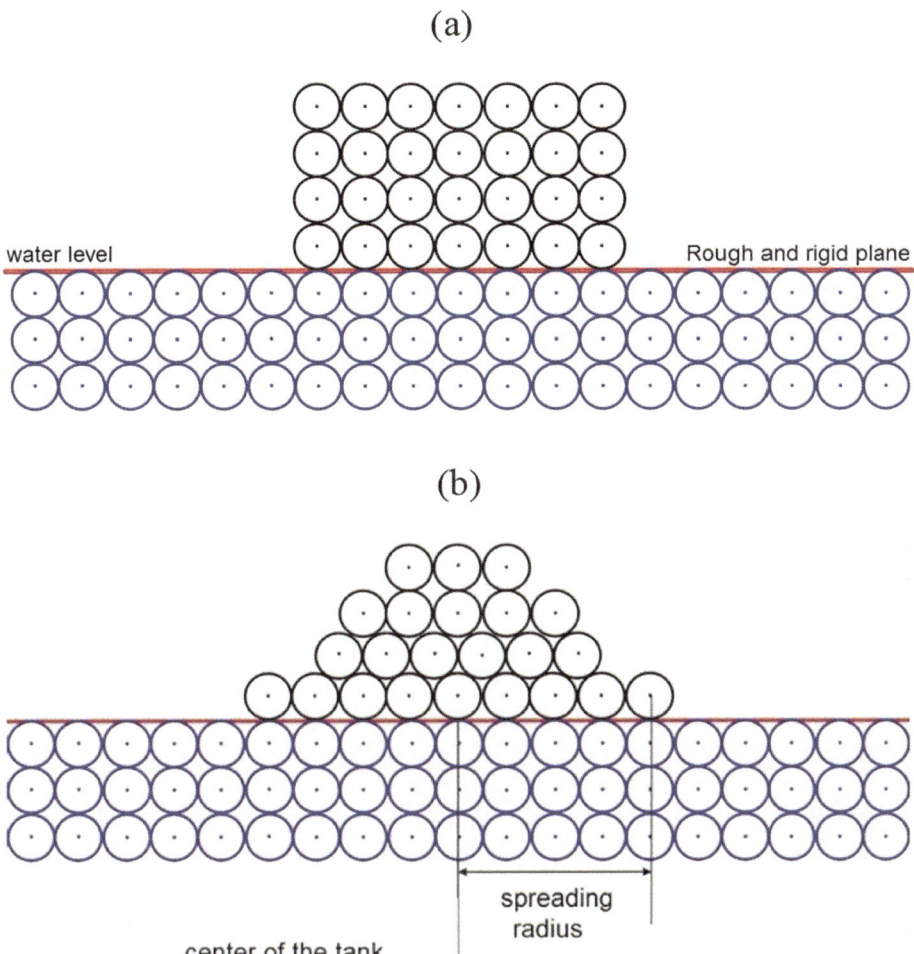

Fig. 3.5 Schematic illustration showing oil and water particles (in black and blue, respectively), separated by a rough and rigid plane. **a** at the initial time instant and **b** sometime after the beginning of oil spreading

3.3.1   Results and Discussions

Simulations were performed using the physical–mathematical model based on Stolzenbach's Hypothesis and different combinations of coefficients of restitution of kinetic energy and friction to calibrate it.

Fay's equation (Eqs. (3.3)) predicted the oil slick diameter at the instant 0.82 s (end of the gravity-inertial spreading) as 3.099 m. Table 3.1 presents the oil slick diameters provided by the SPH/ RBC simulation and the respective per cent error calculated by

Fig. 3.6 Circumferences
illustrate the evolution of the
reach of the oil slick in time.
The inner circumference refers
to the initial diameter of the oil
(at the beginning of the spill)

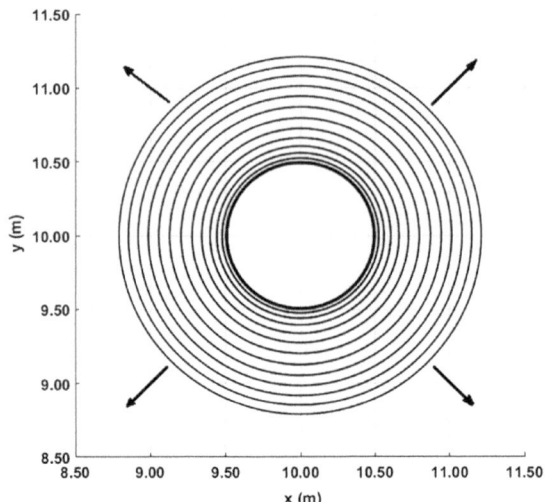

Eq. (3.5) in different definitions of the coefficients of restitution of kinetic energy and friction.

$$\text{Percent error} = \left| \frac{D_{SPH/RBC} - D_{Fay}}{D_{Fay}} \right| \times 100 \qquad (3.5)$$

where $D_{SPH/RBC}$ is the oil slick diameter provided by SPH/ RBC simulation and D_{Fay} is the oil slick diameter predicted by Fay's equation.

Figure 3.7 depicts the evolution of the diameter of the oil slick obtained from the SPH/ RBC modelling (upper curves) for different combinations of coefficients. The lower curves are Fay's equation solution.

At the first instant of the oil spill, the resistant effect of inertia can be seen (the null declivity in the SPH curves). However, the gravitational force prevails, with its driven effect on the spreading predominating over the effect of inertia.

Table 3.1 The oil slick diameters at the end of the gravity-inertial spreading (provided by the SPH/ RBC simulations)

CR	CF	$D_{SPH/RBC}$ (m)	Percent error (%)
0.85	0.15	2.908	6.16
0.85	0.10	3.067	1.03
0.90	0.15	2.915	5.94
0.90	0.10	3.071	0.94
0.92	0.15	2.933	5.36
0.92	0.10	3.126	0.90

Throughout the spreading, the oil slick diameters predicted by SPH/ RBC and Fay diverged and decreased with the calibration of the model (using adequate values for the coefficients CR and CF).

Based on the analysis of the results, the dam-breaking model used for the first instants of the oil spreading (Stolzenbach's Hypothesis) [15] effectively predicted the diameter of the oil slick when the reflective treatment of the contours was utilised. It was emphasised that it is necessary to calibrate the model by choosing the appropriate coefficients, CR and CF.

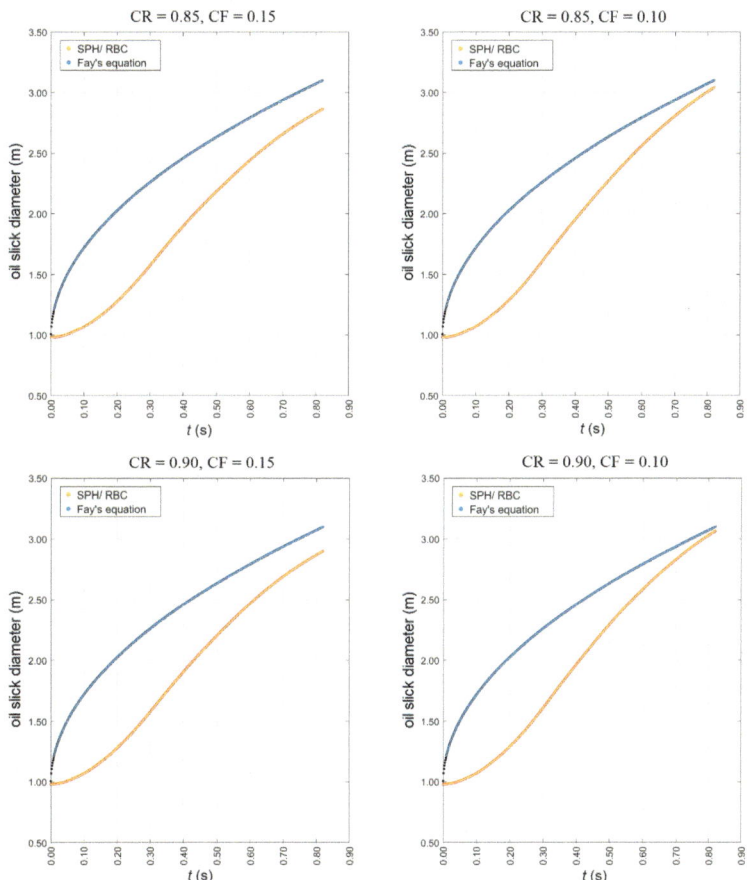

Fig. 3.7 SPH/ RBC simulation results for different combinations of CR and CF

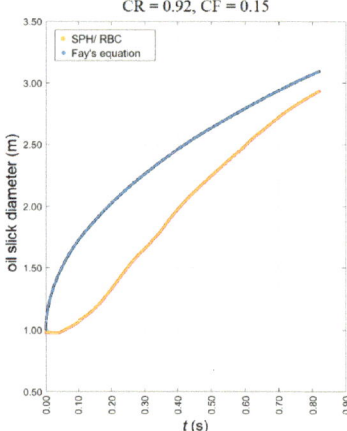
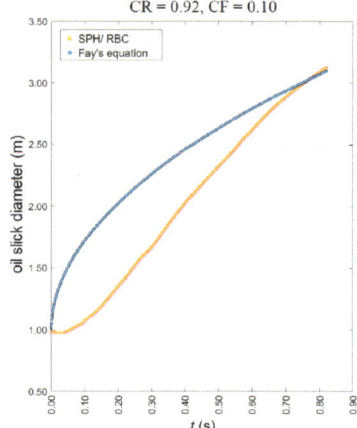

Fig. 3.7 (continued)

References

1. Fourtakas, G., Vacondio, R., Rogers, B. D. On the approximate zeroth and first-order consistency in the presence of 2-D irregular boundaries in SPH obtained by the virtual boundary particle methods. Int. J. Numer. Methods Fluids 78(8), 475–501 (2015)
2. Korzilius S. P., Kruisbrink, A. C. H., Schilders, W. H. A., Anthonissen, M .J.H., Yue, T. Momentum conserving methods that reduce particle clustering in SPH. CASA-Report 2014–2015, Eindhoven: Technische Universiteit Eindhoven. https://pure.tue.nl/ws/files/3858217/376670351 851652.pdf (2014). Accessed July 07, 2024
3. Vorobyev, A. A Smoothed Particle Hydrodynamics Method for the Simulation of Centralized Sloshing Experiments. KIT Scientific Publishing, Germany (2013)
4. Liu M. B., Liu G. R. Smoothed Particle Hydrodynamics (SPH): an Overview and Recent Developments. Arch. Comput. Methods Eng, 17: 25–76 (2010).
5. Vaughan G. L., Healy T. R., Bryan K. R., Sneyd A. D., Gorman R. M. Completeness, conservation and error in SPH for fluids. International Journal for Numerical Methods in Fluids, (56) 37–62 (2008).
6. Korzilius S. P., Kruisbrink A. C. H., Schilders W. H. A., Anthonissen M.J.H.,Yue T. Momentum conserving methods that reduce particle clustering in SPH. CASA-Report 2014–2015, (Technische Universiteit Eindhoven, Eindhoven, 2014), p. 17, available at https://pure.tue.nl/ws/files/3858217/376670351851652.pdf, accessed April 30, 2017.
7. Fourtakas G., Vacondio R., Rogers B. D. On the approximate zeroth and first-order consistency in the presence of 2-D irregular boundaries in SPH obtained by the virtual boundary particle methods. International Journal for Numerical Methods in Fluids, 78(8):475-501 (2015).
8. Batchelor, G. K. An Introduction to Fluid Dynamics, 3rd edn. Cambridge University Press, UK (2000).
9. Monaghan J. J. Simulating free surface flows with SPH. Journal of Computational Physics 110 (2) 399-406 (1994).
10. Newman J. N. Marine Hydrodynamics. MIT Press, Boston (1977).

11. Cruchaga M. A., Celentano D. J., Tezduyar T. E. Collapse of a liquid column: numerical simulation and experimental validation. Computational Mechanics 39, 453-476 (2007).
12. Fraga Filho, C. A. D. An algorithmic implementation of physical reflective boundary conditions in particle methods: Collision detection and response. Physics of Fluids 29, 113602 (2017). https://doi.org/10.1063/1.4997054
13. Fay J.A. The Spread of Oil Slicks on a Calm Sea. Oil on the Sea, Plenum Press (1969) 53-64.
14. Fay J.A., Physical Processes in the Spread of Oil on a Water Surface, International Oil Spill Conference Proceedings (1971) 463–467.
15. Stolzenbach K.D., Madsen O.S., Adams E.E., Pollack A.M., Cooper C.K. A Review and Evaluation of Basic Techniques for Predicting the Behavior of Surface Oil Slicks. Massachusets Institute of Technology, (1977).
16. Fraga Filho C.A.D. A Lagrangian analysis of the gravity-inertial oil spreading on the calm sea using the reflective oil-water interface treatment. Environ Sci Pollut Res. (2021). https://doi.org/10.1007/s11356-020-11508-2

Applications of Reflective Boundary Conditions in Three-Dimensional Domains

4

The more recent implementations, validation and discussion of results of Reflective Boundary Conditions coupled with the SPH method in the three-dimensional domain will be presented in this chapter.

4.1 Uniform and Incompressible Fluid at Rest Inside a Reservoir

This hydrostatics problem has been solved in two dimensions (as shown in Sect. 3.1) and, using the modified pressure concept, was also solved for the three-dimensional domain.

The domain consists of a reservoir open to the atmosphere, filled with a Newtonian, incompressible, uniform and isothermal liquid. The dimensions of the tank are $1.0 \text{ m} \times 1.0 \text{ m} \times 1.0 \text{ m}$. The water particles inside the reservoir are at 20 °C and at sea level ($\rho = 1.00 \times 10^3 \text{ kg/m}^3$, ν_{water} $\nu = 1.00 \times 10^{-6} \text{ m}^2/\text{s}$). Twenty-five particles per side of the tank (15,625 in total) were used to discretise of the fluid. The physical laws of conservation of mass and momentum in the continuum domain have been employed in problem-solving [1]. The time step used was 1.00×10^{-4} s. The modified pressure concept has been used (Eq. (3.1)), changing the form of the momentum conservation equation (Eq. (3.2)).

The simulations were run for a simulated time of 5.00 s (50,000 numerical iterations). Figure 4.1a shows the centres of mass of the fluid particles inside the reservoir and the hydrostatic pressure field in the initial time instant. Figure 4.1b depicts the centres of mass of the fluid particles and the modified pressure field in the initial instant.

A zero coefficient of restitution of kinetic energy was used, and the particle velocities' responses were also zero. In the same way as solving this problem in a two-dimensional

© The Author(s), under exclusive license to Springer Nature Switzerland AG 2025
C. A. D. Fraga Filho, *Reflective Boundary Conditions in SPH Fluid Dynamics Simulation*, Synthesis Lectures on Mechanical Engineering,
https://doi.org/10.1007/978-3-031-71582-2_4

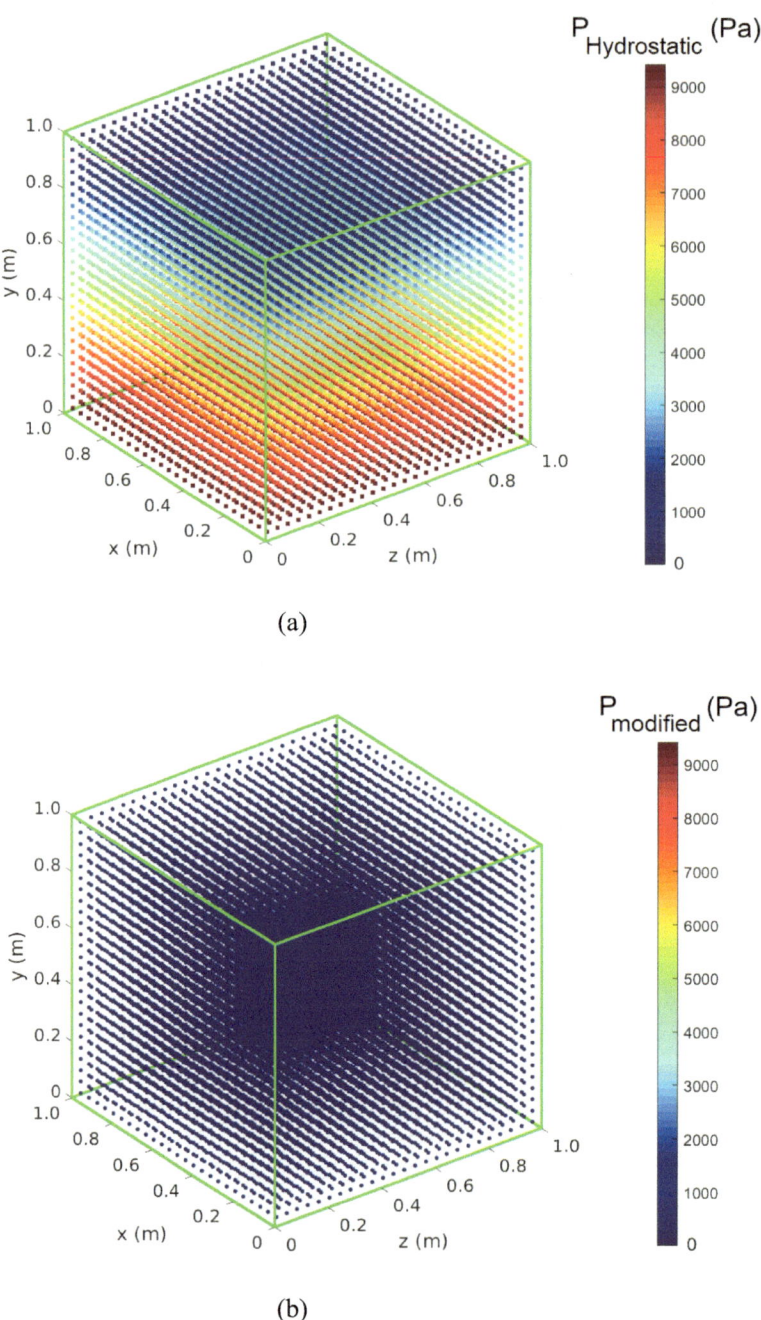

Fig. 4.1 Pressure fields acting on the particles inside the tank. **a** The hydrostatic pressure field. **b** The modified pressure field (equal to 0.0 Pa). The particles are represented by their centres of mass. Reproduced from [1], with the permission of AIP Publishing

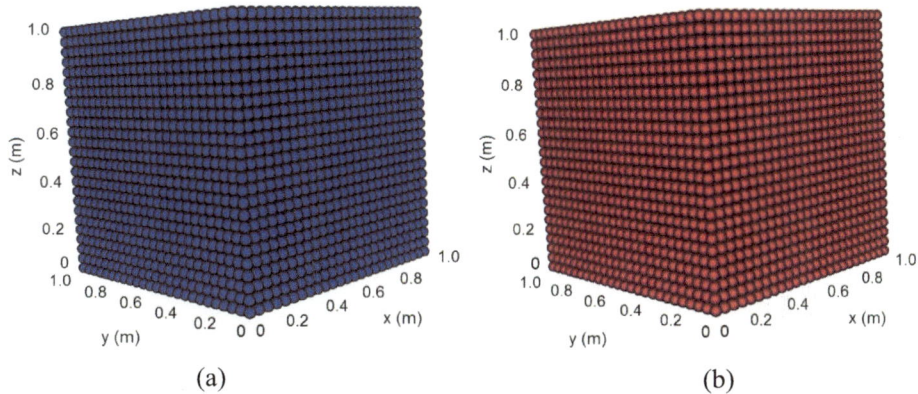

Fig. 4.2 The water particles at the initial time instant (in blue) and at the final instant of the simulation 5.00 s (in red). Reproduced from [1], with the permission of AIP Publishing

domain, the CDRA detected the contact between fluid particles and the lateral and lower planes, i.e. the distance between the centres of mass of those particles and the plane was equal to the particle radius. During the response stage, particles' reflections were carried out based on Eq. (2.9) with no change in the positions of the centres of mass.

The coincidence of the positions of the centres of mass in the final and initial instants of the simulation (5.00 s/ 50,000 iterations) was verified, as shown in Fig. 4.2.

4.2 Dam-Breaking Over a Dry Bed

As in the two-dimensional problem in Sect. 3.2, the fluid is assumed to be incompressible, uniform, and isothermal. The solution of the Navier–Stokes equations, mass and momentum conservation, provided the particles' acceleration, and shortly after, their positions and velocities (input data for the CDRA) were obtained through temporal integration. Figure 4.3 depicts the simulated computational domain and the initial particle setup.

The water damned in the reservoir measured 0.420 m in length, 0.440 m in height and 0.228 m in depth, with a volume of 5.926×10^{-3} m^3, which was discretised by 32,000 particles, with initial lateral distances of 5.70 mm between two consecutive centres of mass. The physical properties of water at a temperature of 20 °C were a density of 1.00×10^3 kg/m^3 and a kinematic viscosity of 1.00×10^{-6} m^2/s.

The absolute pressure of the fluid was calculated as the sum of a hydrostatic and a dynamic component (based on Tait's state equation) [2]. The particles on the free surface were identified, and a constant pressure condition of zero was applied and updated at each numerical iteration using Newman boundary conditions [3]. A sub-grid scale (SGS) model has been employed to treat the turbulence. There was no significant difference

Fig. 4.3 The geometry
simulated and the damned
water (32,000 particles were
used in the discretisation of the
fluid). Reproduced from [1],
with the permission of AIP
Publishing

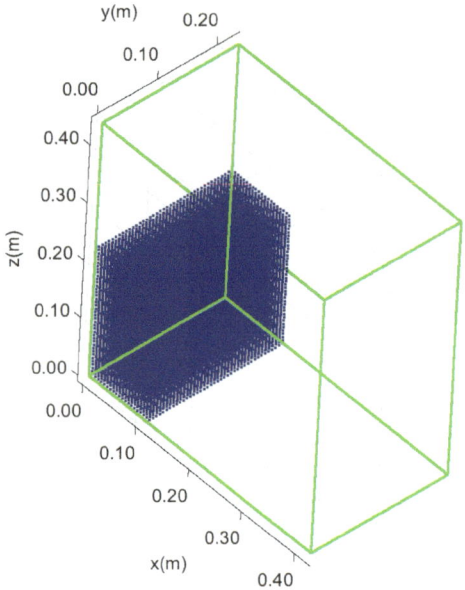

in the numerical results (when compared with the laminar shear stresses model used in
simulation) [4]. A time step of 1.00×10^{-4} s was used in the simulation.

After the solution of physical conservation equations of mass and momentum by the
Lagrangian SPH Method, the 1st-order Runge–Kutta integration method (Euler's method)
has been applied to update the positions and velocities of the particles in each numerical
iteration and provide input data for the CDRA (in which the coefficients CR and CF
received the values 1.00 and 0.00, respectively). The simulations were run for a physical
simulated time of 1.0 s (10,000 numerical iterations). The time evolution of the dam-break
flow is depicted in Fig. 4.4.

Figure 4.5 depicts the free surface profiles (experimental, provided by FEM and SPH/
RBC).

Based on the analysis of Fig. 4.5, a good agreement between the simulation results
(SPH/ RBC), the experimental data and, mainly, the FEM results was observed.

The areas under the free surfaces provided by the SPH method coupled with reflective
boundary conditions (SPH/ RBC) and the FEM were calculated. The initial water area at
the plane xz is 2.540×10^{-2} m^2 (length: 0.114 m, height: 0.228 m). Table 4.1 displays
the areas under the free surface lines from FEM and SPH method coupled with reflective
boundary conditions, with percentage differences in Table 4.2.

Tables 4.1 and 4.2 display the areas under the free surface as calculated by the FEM
and SPH/ RBC methods. The most significant difference between the initial water area (at
the xz plane) and the area obtained from the SPH/RBC simulation was observed at the end

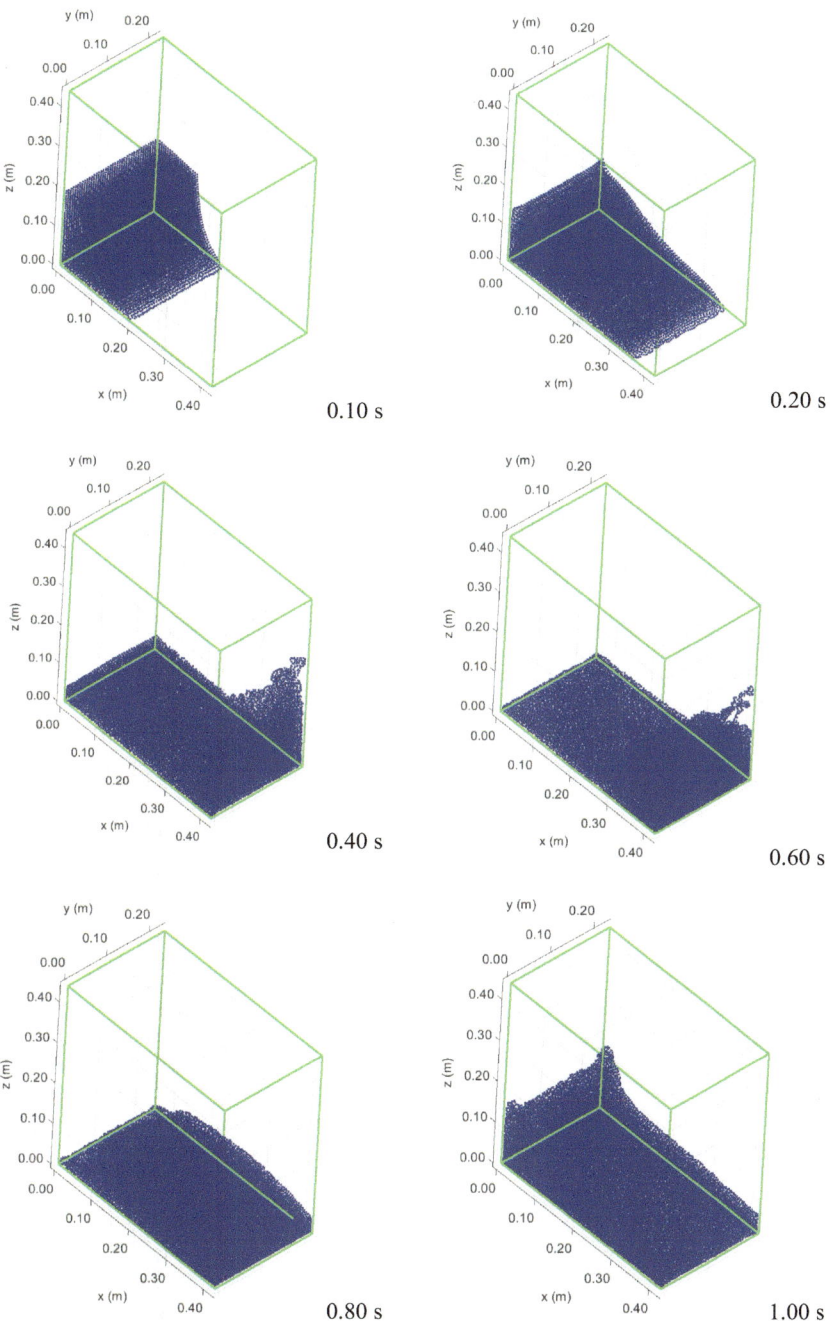

Fig. 4.4 Dam-break flow evolution (results provided by the SPH Method coupled with reflective boundary conditions). Reproduced from [1], with the permission of AIP Publishing

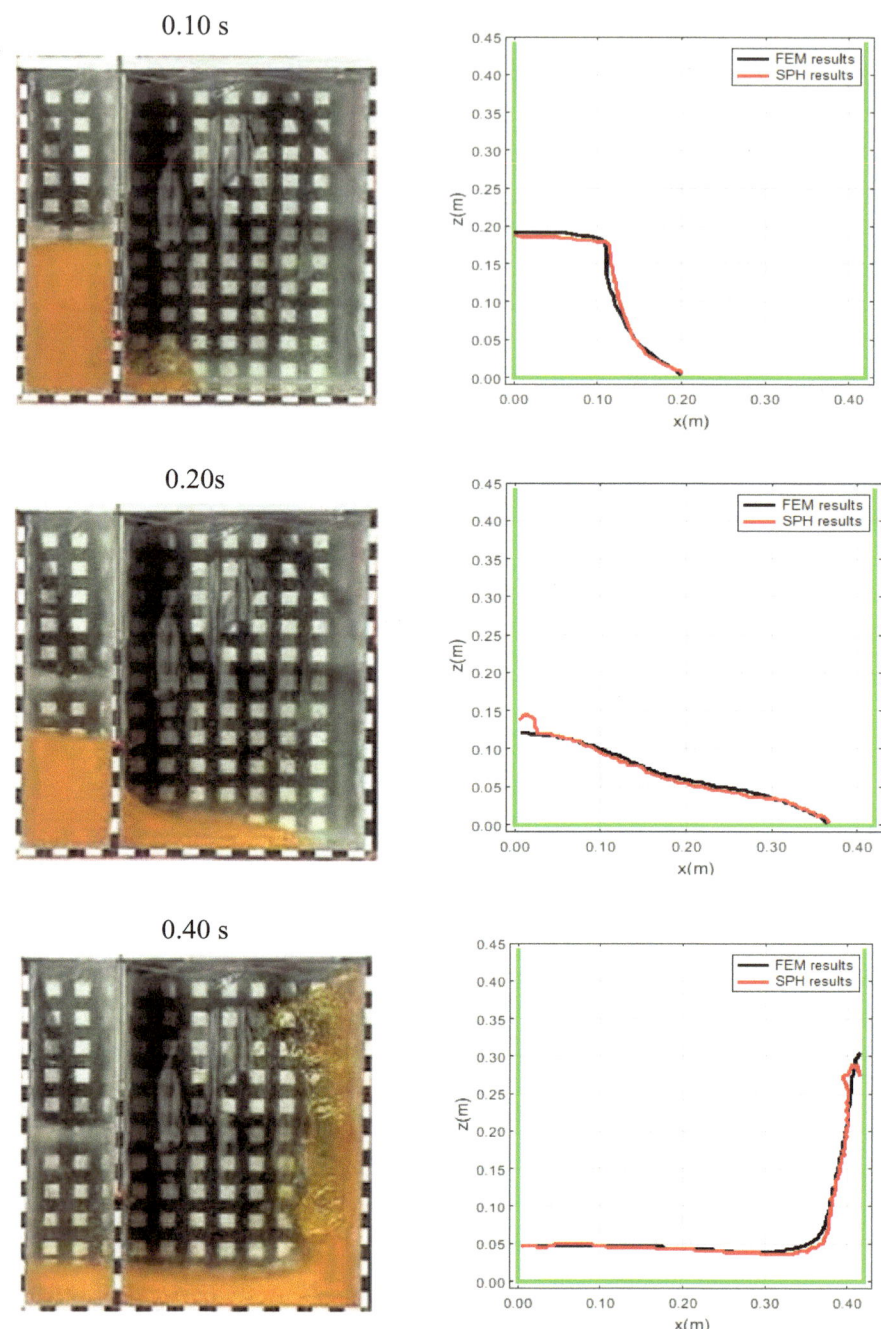

Fig. 4.5 Experimental dam-break flow results in the first column (section xz) and free surface profiles provided by Finite Element Method (FEM), in black, and SPH Method coupled with reflective boundary conditions, in red, in the second column. Reproduced from [1], with the permission of AIP Publishing

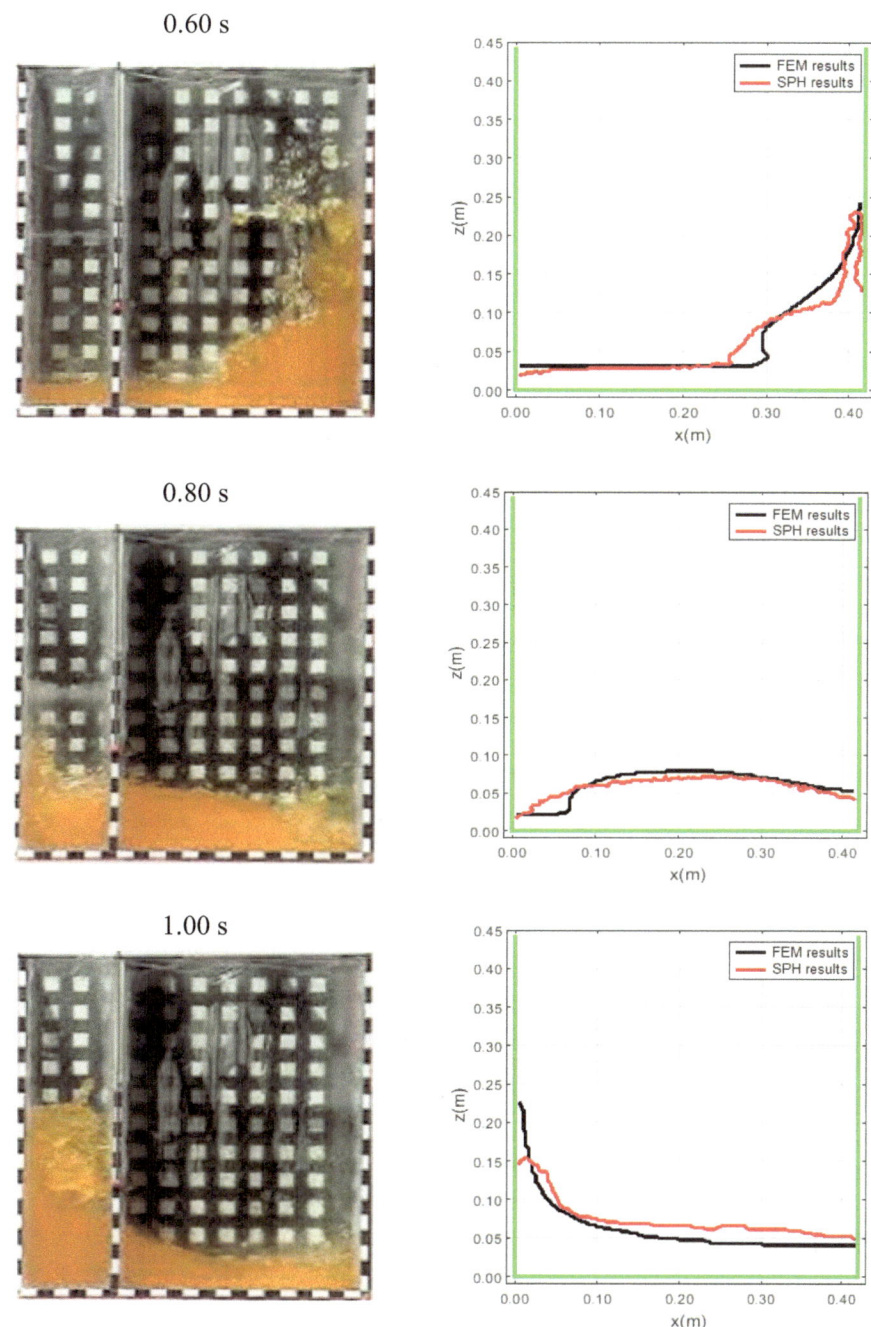

Fig. 4.5 (continued)

Table 4.1 Areas under free surfaces lines provided by FEM and SPH method coupled with reflective boundary conditions ($\times 10^{-2}$ m^2)

Method	Time (s)									
	0.10	0.20	0.30	0.40	0.50	0.60	0.70	0.80	0.90	1.00
FEM	2.52	2.40	2.52	2.47	2.41	2.46	2.52	2.55	2.53	2.49
SPH/ RBC	2.51	2.43	2.35	2.09	2.35	2.44	2.35	2.43	2.71	3.04

Table 4.2 Percentual differences in relation to the initial area occupied by the water (plane xz)

Method	Time (s)									
	0.10	0.20	0.30	0.40	0.50	0.60	0.70	0.80	0.90	1.00
FEM	−0.78	−5.51	−0.79	−2.76	−5.12	−3.15	−0.79	0.39	−0.39	−1.97
SPH/ RBC	−1.18	−4.33	−7.48	−17.72	−7.48	−3.94	−7.48	−4.33	6.69	19.69

of the simulation (1.00 s), amounting to 19.69%. More complete results and information can be found in [1].

4.3   Fluid–Structure Interaction

Fluid–structure interaction (FSI) is a common phenomenon in daily life, including natural systems, engineering, aerospace, and medicine. FSI can be observed in wave impacts against reservoir walls, coastal structures such as offshore platforms and bridge piers, sloshing phenomena, tsunamis, coastal flooding, and biomedical applications.

A benchmark case study, previously investigated in the literature [5–7], was simulated using SPH/RBC. It consisted of a three-dimensional dam-break flow over a dry bed, generating a wave that impacted a tall structure (a rigid obstacle) fixed inside the reservoir. Dynamic particles were used to treat the contours in these simulations [8, 9]. The simulated geometry and initial particle setup are shown in Fig. 4.6.

The reservoir had the following dimensions: 160 cm (length) × 61 cm (width) × 75 cm (height). Initially, the volume of water damned in the reservoir was 40 cm (length) × 61 cm (width) × 30 cm (height). Inside the tank was a rigid obstacle in the form of a tall structure. This structure was defined by the intersection of six geometric planes: four vertical faces (sides), one upper and one lower. It measured 12 cm × 12 cm × 75 cm and was positioned 50 cm downstream of the gate (located at x = 40 cm) and 24 cm away from the closest side wall of the tank. 12,716 Lagrangian particles (34, 22, 17, in the x, y and z directions, respectively) were used to discretise the fluid. The distance between the centres of mass of two adjacent particles was 1.79×10^{-2} m.

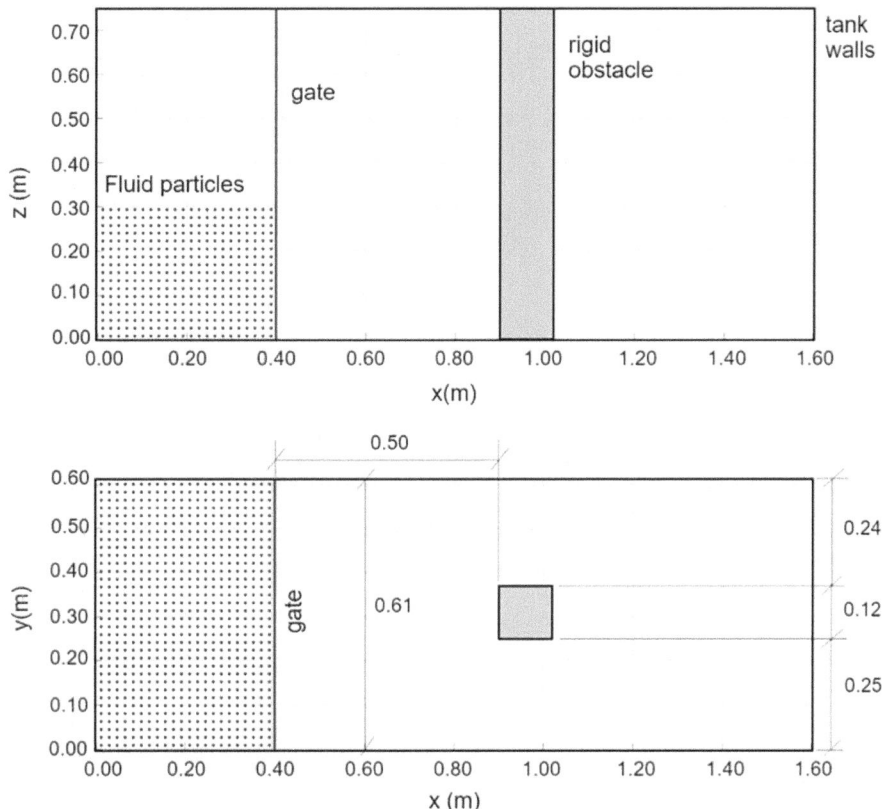

Fig. 4.6 The geometry of the reservoir and the fixed obstacle (lateral and upper views) is shown above, along with the initial volume of dammed water discretised by particles. Reproduced from [10], with the permission of Springer Nature

The physical properties of water at 20 °C were a density (ρ) of 1.00×10^3 kg/m³ and a kinematic viscosity (ν) of 1.00×10^{-6} m²/s. The pressure acting on the free surface was the atmospheric pressure (taken as a reference). The particles defining the free surface at the initial time were marked, and their pressures were set to zero (Newman boundary conditions) and remained constant throughout the numerical simulation [3]. The fluid's absolute pressure was calculated as the sum of a hydrostatic and a dynamic component, as predicted by Tait's state equation [2]. A model to capture the turbulence effects was not used in the simulation. The coefficients of restitution for kinetic energy (CR) and friction (CF) in the CDRA were 1.00 and 0.00, respectively.

Figure 4.7 shows the evolution of the particles' centres of mass and the magnitude of their velocities in some time instants. The colours were defined according to the classes of the fluid velocities: blue: [0.00, 0.53) m/s; green: [0.53, 1.06) m/s; yellow: [1.06, 1.59) m/s, and red above 1.59 m/s.

Numerical validated results presented in the literature [7] (Fig. 4.8) were used to analyse the SPH/ RBC simulation results.

The SPH/ RBC simulation results (Fig. 4.7) showed good conformity with the literature results (Fig. 4.8).

In Fig. 4.9, the xz lateral cross-section evolution of the particles is shown for the simulation conducted in this work (line 1) and provided by the literature [5] (line 2) in

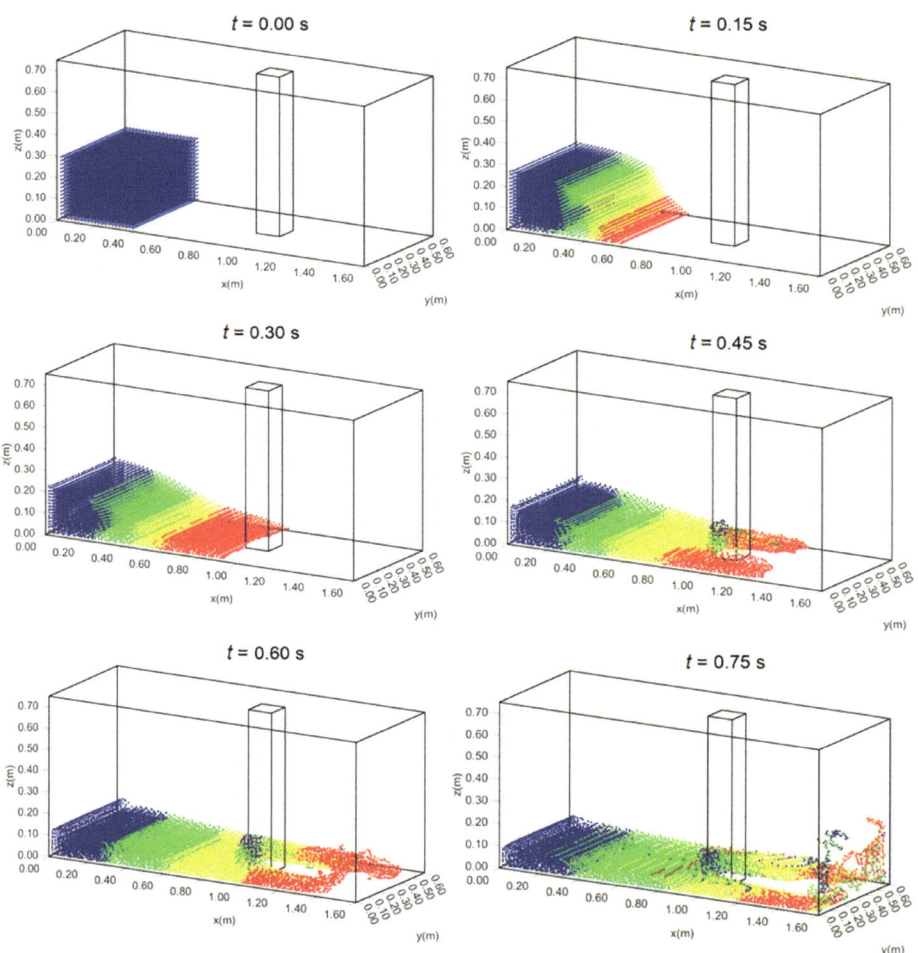

Fig. 4.7 Graphical visualisation of SPH/ RBC simulation results in some time instants. The colours are associated with the magnitudes of the particles' velocities. Reproduced from [10], with the permission of Springer Nature

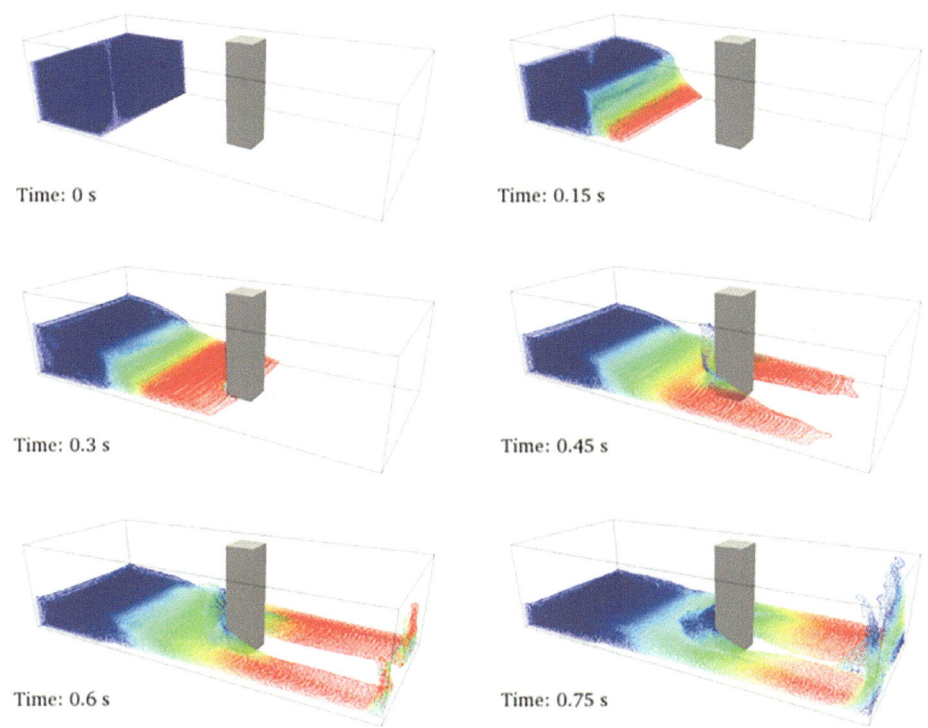

Time: 0 s

Time: 0.15 s

Time: 0.3 s

Time: 0.45 s

Time: 0.6 s

Time: 0.75 s

Fig. 4.8 Simulation results provided by [7]. With permission of the authors

two different time instants. The waveforms in both simulations appear similar. At the time instant 0.21 s, the wavefront positions were 0.71 m and 0.73 m (from the literature), resulting in a percentage difference of 2.74%. At 0.27 s, the wavefront positions were 0.86 m (line 1) and 0.90 m (line 2), leading to a percentage difference of 4.44%.

The highest heights of the wave at 0.21 s were 0.26 m (in this work) and 0.23 m (in the literature [5]), with a percentage difference of 3.00%. In 0.27 s, they were 0.23 m and 0.21 m, respectively, with a percentage difference of 9.00%.

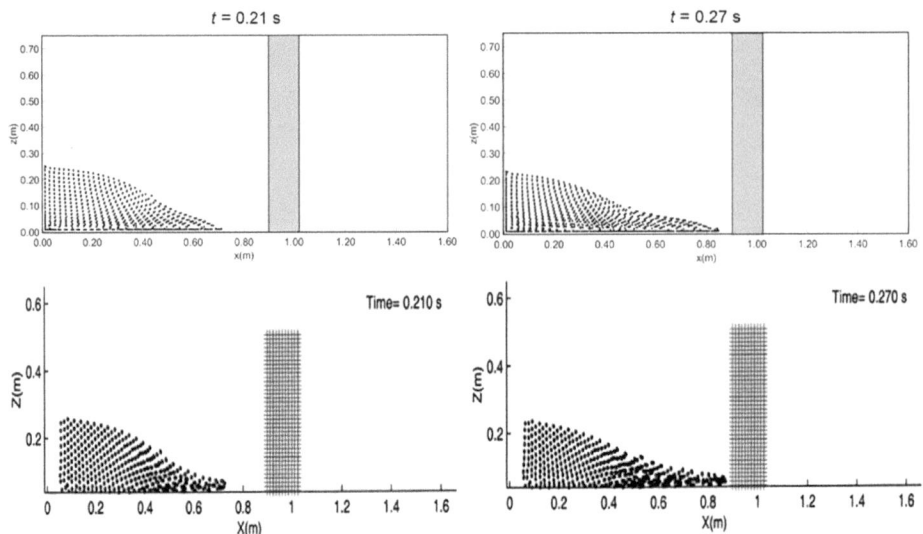

Fig. 4.9 In the first line, the lateral xz cross-section shows the evolution of the fluid particles in this work's simulation. The SPH/ RBC simulation results showed a good agreement with the literature results, validated from experiments. Reproduced from [5], with the permission from ASCE

References

1. Fraga Filho, C. A. D., Peng C., Islam R. I., McCabe C., Baig B., Venkata Durga Prasad, G. V. D. Implementation of three-dimensional physical reflective boundary conditions in mesh-free particle methods for continuum fluid dynamics: Validation tests and case studies. Phys. Fluids 31, 103606 (2019). https://doi.org/10.1063/1.5115776
2. Monaghan J. J. Simulating free surface flows with SPH. Journal of Computational Physics 110 (2) 399-406 (1994).
3. Newman J. N. Marine Hydrodynamics. MIT Press, Boston (1977).
4. Filho, C. A. D. F., Piccoli, F. P. Diffusive terms applied in smoothed particle hydrodynamics simulations of incompressible and isothermal Newtonian fluid flows. J Braz. Soc. Mech. Sci. Eng. 43, 479 (2021). https://doi.org/10.1007/s40430-021-03158-3
5. Gómez-Gesteira M., Dalrymple R. A. Using a Three-Dimensional Smoothed Particle Hydrodynamics Method for Wave Impact on a Tall Structure. Journal of Waterway, Port, Coastal and Ocean Engineering (2004). https://doi.org/10.1061/(ASCE)0733-950X(2004)130:2(63)
6. Domínguez J. M. DualSPHysics: Towards High Performance Computing using SPH technique, PhD Thesis, Universidade de Vigo, Spain. Available at https://www.researchgate.net/public ation/268216668_DualSPHysics_Towards_High_Performance_Computing_using_SPH_techni que/link/54651a450cf2052b509f2c17/download (2014). Accessed on 15 September, 2023.
7. Crespo A. J. C. , Domínguez J. M., Rogers B. D., Gómez-Gesteira M., Longshaw S., Canelas R., Vacondio R., Barreiro R. A., García-Feal O. DualSPHysics: Open-source parallel CFD solver

based on Smoothed Particle Hydrodynamics (SPH). Comput. Phys. Commun (2015). https://doi.org/10.1016/j.cpc.2014.10.004

8. Crespo, A.J.C., Gómez-Gesteira, M., Dalrymple, R.A. (2007) Boundary conditions generated by dynamic particles in SPH methods. CMC Comput. Mat. Cont. 5(3), 173–184. https://doi.org/10.3970/cmc.2007.005.173

9. Fraga Filho, C.A.D. Smoothed particle hydrodynamics fundamentals and basic applications in continuum mechanics. Springer Nature, Switzerland (2019).

10. Fraga Filho, C.A.D. Reflective boundary conditions coupled with the SPH method for the three-dimensional simulation of fluid–structure interaction with solid boundaries. J Braz. Soc. Mech. Sci. Eng. 46, 256 (2024). https://doi.org/10.1007/s40430-024-04807-z

Conclusions

Over the past ten years, Reflective Boundary Conditions have been implemented and validated for applications in hydrostatics and hydrodynamics problems (in two- and three-dimensional domains).

The primary motivation for developing scientific research in Reflective Boundary Conditions was the search to replace artificial techniques still widely used in Meshfree Lagrangian Particle Methods with physical and realistic strategies for treating contours, which respect the continuum laws.

The numerical results obtained using the SPH Lagrangian Method and the Reflective Boundary Conditions in the hydrostatic cases. They strongly agreed with the results of the literature on hydrodynamic cases presented in this book. This agreement encourages the continuity of applying the reflective boundary conditions for solving other scientific problems in 2D and 3D domains.

© The Editor(s) (if applicable) and The Author(s), under exclusive license   45
to Springer Nature Switzerland AG 2025
C. A. D. Fraga Filho, *Reflective Boundary Conditions in SPH Fluid Dynamics Simulation*, Synthesis Lectures on Mechanical Engineering,
https://doi.org/10.1007/978-3-031-71582-2

Appendix A
CDRA Validation in Two-Dimensional Domains

Figure A.1 shows a longitudinal section of a closed reservoir defined by four perpendicular planes, and a particle in its initial position. The collisions of a single particle against these planes have been verified by a comparison of the final positions and velocities of its centre of mass, that is, those provided by the algorithm at the end of each time step, and those obtained from mathematical modelling (analytical results).

Tests were carried out in which the particle radius was 0.01 m, the height and the length of the reservoir were 0.50 m, and the acceleration was zero; the objective was to verify the detection and the response of the collisions, and thus the acceleration was not considered in the algorithm. The initial position and velocity of the centre of mass was varied in these tests. The Euler method was used for temporal integration, with a time step of 1.00 s, and the coefficient of restitution of energy was varied. Some of the results obtained are presented below.

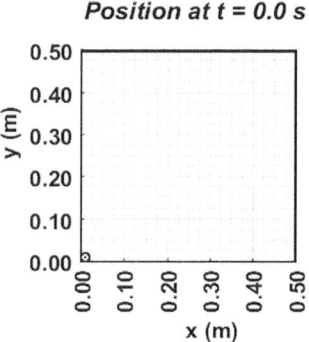

Fig. A.1 Box and particle in its initial position. Reproduced from Fraga Filho, C. A. D. An algorithmic implementation of physical reflective boundary conditions in particle methods: Collision detection and response. Physics of Fluids 29, 113,602 (2017), with the permission of AIP

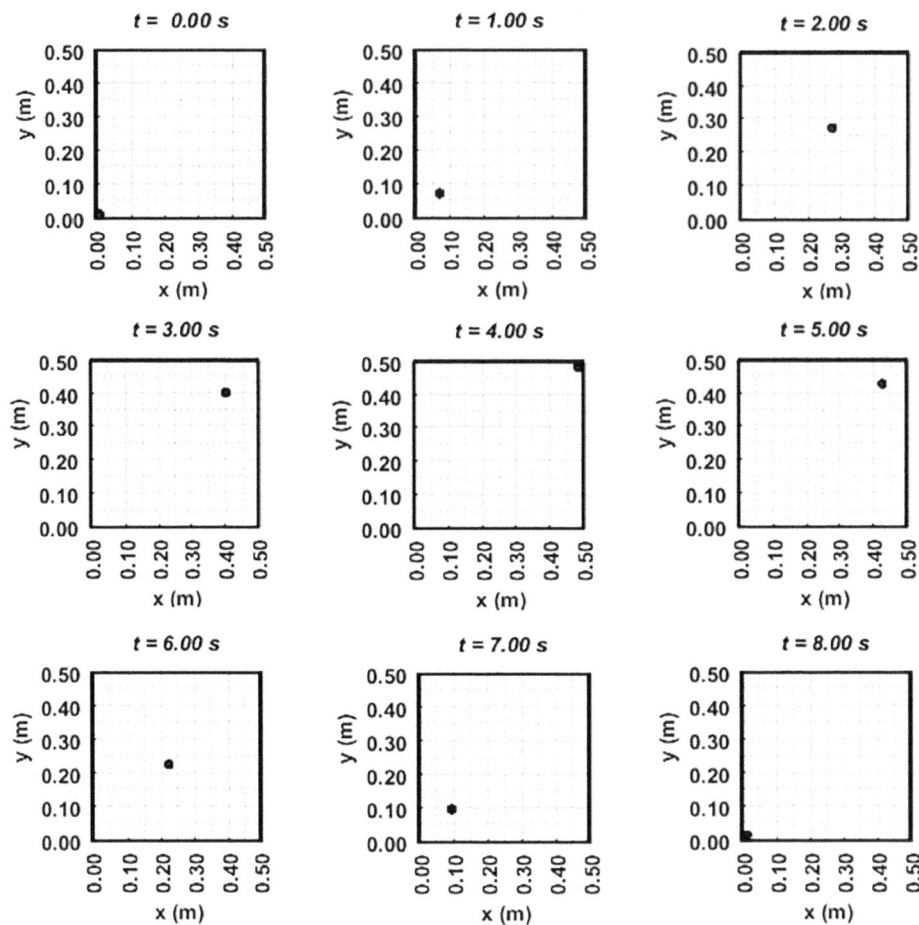

Fig. A.2 Positions of the centre of mass in the first eight seconds of the 1st test

Test 1:

Initial position of the centre of mass: $C_o = (0.01, 0.01)$ m.

Initial velocity: $V_o = (1.00, 1.00)$ m/s.

CR $= 1.00$ (Fig. A.2 and Table A.1).

Test 2:

Initial position of the centre of mass: $C_o = (0.25, 0.25)$ m.

Initial velocity: $V_o = (0.10, 0.00)$ m/s.

CR $= 1.00$ (Fig. A.3 and Table A.2).

Table A.1 Numerical results of the first validation test (SI units)*

t	Input data						Output data			
	V_{ox}	V_{oy}	C_{ox}	C_{oy}	C_{1x}	C_{1y}	C_{fx}	C_{fy}	V_{fx}	V_{fy}
1.00	1.00	1.00	0.01	0.01	1.01	1.01	0.05	0.05	1.00	1.00
2.00	1.00	1.00	0.05	0.05	1.05	1.05	0.09	0.09	1.00	1.00
3.00	1.00	1.00	0.09	0.09	1.09	1.09	0.13	0.13	1.00	1.00
4.00	1.00	1.00	0.13	0.13	1.13	1.13	0.17	0.17	1.00	1.00
5.00	1.00	1.00	0.17	0.17	1.17	1.17	0.21	0.21	1.00	1.00
6.00	1.00	1.00	0.21	0.21	1.21	1.21	0.25	0.25	1.00	1.00
7.00	1.00	1.00	0.25	0.25	1.25	1.25	0.29	0.29	1.00	1.00
8.00	1.00	1.00	0.29	0.29	1.29	1.29	0.33	0.33	1.00	1.00

* Rounded to two decimal places

Test 3:

Initial position of the centre of mass: $C_o = (0.25, 0.25)$ m.

Initial velocity: $V_o = (0.00, 1.00)$ m/s.

CR $= 1.00$ (Fig. A.4 and Table A.3).

Test 4:

Initial position of the centre of mass: $C_o = (0.01, 0.01)$ m.

Initial velocity: $V_o = (0.87, 0.50)$ m/s.

CR $= 1.00$ (Fig. A.5 and Table A.4).

Test 5:

Initial position of the centre of mass: $C_o = (0.01, 0.01)$ m.

Initial velocity: $V_o = (1.00, 1.00)$ m/s.

CR $= 0.90$ (Fig. A.6 and Table A.5).

Test 6:

Initial position of the centre of mass: $C_o = (0.25, 0.25)$ m.

Initial velocity: $V_o = (0.10, 0.00)$ m/s.

CR $= 0.90$ (Fig. A.7 and Table A.6).

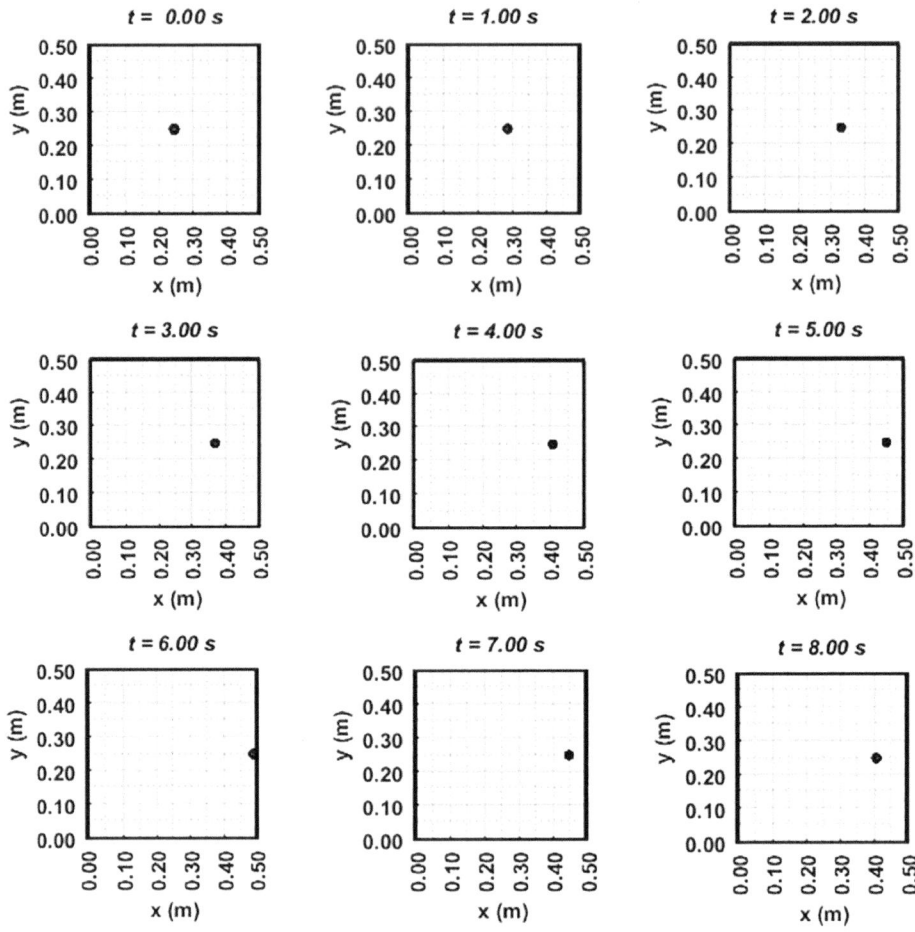

Fig. A.3 Positions of the centre of mass in the first eight seconds of the 2nd test

Test 7

Initial position of the centre of mass: $C_o = (0.25, 0.25)$ m.

Initial velocity: $V_o = (0.00, 1.00)$ m/s.

$CR = 0.90$ (Fig. A.8 and Table A.7).

Table A.2 Results of the second validation test (SI units)*

t	Input data						Output data			
	V_{ox}	V_{oy}	C_{ox}	C_{oy}	C_{1x}	C_{1y}	C_{fx}	C_{fy}	V_{fx}	V_{fy}
1.00	1.00	0.00	0.25	0.25	1.25	0.25	0.29	0.25	1.00	0.00
2.00	1.00	0.00	0.29	0.25	1.29	0.25	0.33	0.25	1.00	0.00
3.00	1.00	0.00	0.33	0.25	1.33	0.25	0.37	0.25	1.00	0.00
4.00	1.00	0.00	0.37	0.25	1.37	0.25	0.41	0.25	1.00	0.00
5.00	1.00	0.00	0.41	0.25	1.41	0.25	0.45	0.25	1.00	0.00
6.00	1.00	0.00	0.45	0.25	1.45	0.25	0.49	0.25	−1.00	0.00
7.00	−1.00	0.00	0.49	0.25	−0.51	0.25	0.45	0.25	−1.00	0.00
8.00	−1.00	0.00	0.45	0.25	−0.55	0.25	0.41	0.25	−1.00	0.00

* Rounded to two decimal places

Test 8:

Initial position of the centre of mass: $C_o = (0.01, 0.01)$ m.

Initial velocity: $V_o = (0.87, 0.50)$ m/s.

CR = 0.90 (Fig. A.9 and Table A.8).

Test 9:

Initial position of the centre of mass: $C_o = (0.01, 0.01)$ m.

Initial velocity: $V_o = (0.10, 0.10)$ m/s.

CR = 0.80 (Fig. A.10 and Table A.9).

Test 10:

Initial position of the centre of mass: $C_o = (0.25, 0.25)$ m.

Initial velocity: $V_o = (0.10, 0.00)$ m/s.

CR = 0.80 (Fig. A.11 and Table A.10).

Test 11:

Initial position of the centre of mass: $C_o = (0.25, 0.25)$ m.

Initial velocity: $V_o = (0.00, 0.10)$ m/s.

CR = 0.80 (Fig. A.12 and Table A.11).

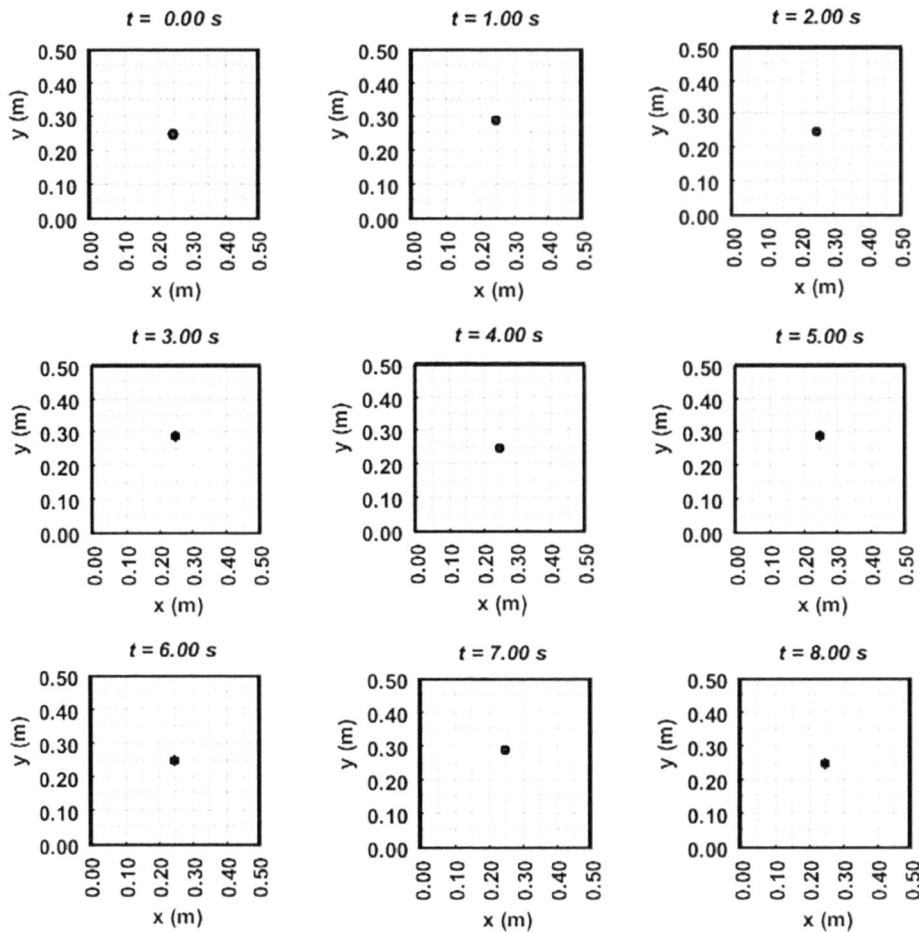

Fig. A.4 Positions of the centre of mass in the first eight seconds of the 3rd test

Test 12:

Initial position of the centre of mass: $C_o = (0.01, 0.01)$ m.

Initial velocity: $V_o = (0.87, 0.50)$ m/s.

CR = 0.80 (Fig. A.13 and Table A.12).

Table A.3 Results of the third validation test (SI units)*

t	Input data						Output data			
	V_{ox}	V_{oy}	C_{ox}	C_{oy}	C_{1x}	C_{1y}	C_{fx}	C_{fy}	V_{fx}	V_{fy}
1.00	0.00	1.00	0.25	0.25	0.25	1.25	0.25	0.29	0.00	−1.00
2.00	0.00	−1.00	0.25	0.29	0.25	−0.71	0.25	0.25	0.00	1.00
3.00	0.00	1.00	0.25	0.25	0.25	1.25	0.25	0.29	0.00	−1.00
4.00	0.00	−1.00	0.25	0.29	0.25	−0.71	0.25	0.25	0.00	1.00
5.00	0.00	1.00	0.25	0.25	0.25	1.25	0.25	0.29	0.00	−1.00
6.00	0.00	−1.00	0.25	0.29	0.25	−0.71	0.25	0.25	0.00	1.00
7.00	0.00	1.00	0.25	0.25	0.25	1.25	0.25	0.29	0.00	−1.00
8.00	0.00	−1.00	0.25	0.29	0.25	−0.71	0.25	0.25	0.00	1.00

* Rounded to two decimal places

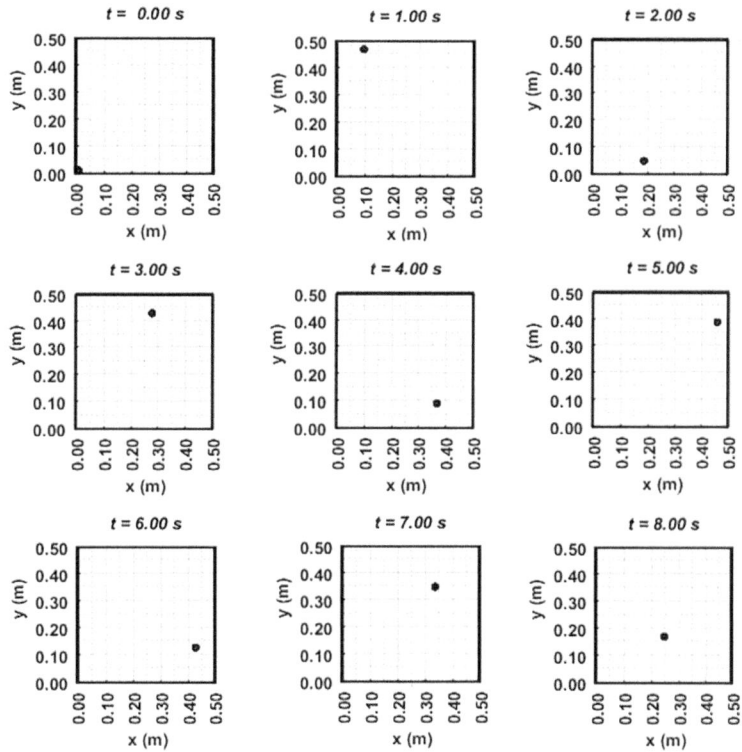

Fig. A.5 Positions of the centre of mass in the first eight seconds of the 4th test

Table A.4 Results of the fourth validation test (SI units)*

t	Input data						Output data			
	V_{ox}	V_{oy}	C_{ox}	C_{oy}	C_{1x}	C_{1y}	C_{fx}	C_{fy}	V_{fx}	V_{fy}
1.00	0.87	0.50	0.01	0.01	0.88	0.51	0.10	0.47	−0.87	−0.50
2.00	−0.87	−0.50	0.10	0.47	−0.77	−0.30	0.19	0.05	−0.87	0.50
3.00	−0.87	0.50	0.19	0.05	−0.68	0.55	0.28	0.43	−0.87	−0.50
4.00	−0.87	−0.50	0.28	0.43	−0.59	−0.07	0.37	0.09	−0.87	0.50
5.00	−0.87	0.50	0.37	0.09	−0.50	0.59	0.46	0.39	−0.87	−0.50
6.00	−0.87	−0.50	0.46	0.39	−0.41	−0.11	0.43	0.13	0.87	0.50
7.00	0.87	0.50	0.43	0.13	1.30	0.63	0.34	0.35	0.87	−0.50
8.00	0.87	−0.50	0.34	0.35	1.21	−0.15	0.25	0.17	0.87	0.50

* Rounded to two decimal places

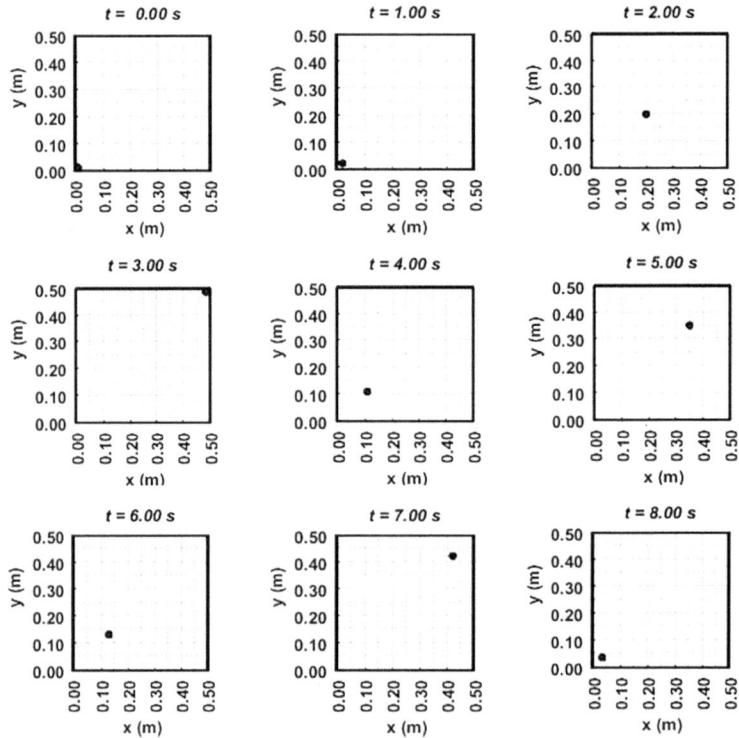

Fig. A.6 Positions of the centre of mass in the first eight seconds of the 5th test

Table A.5 Results of the fifth validation test (SI units)*

t	Input data						Output data			
	V_{ox}	V_{oy}	C_{ox}	C_{oy}	C_{1x}	C_{1y}	C_{fx}	C_{fy}	V_{fx}	V_{fy}
1.00	1.00	1.00	0.01	0.01	1.01	1.01	0.02	0.02	−0.90	−0.90
2.00	−0.90	−0.90	0.02	0.02	−0.88	−0.88	0.20	0.20	−0.73	−0.73
3.00	−0.73	−0.73	0.20	0.20	−0.53	−0.53	0.49	0.49	−0.59	−0.59
4.00	−0.59	−0.59	0.49	0.49	−0.10	−0.10	0.11	0.11	0.53	0.53
5.00	0.53	0.53	0.11	0.11	0.64	0.64	0.35	0.35	−0.48	−0.48
6.00	−0.48	−0.48	0.35	0.35	−0.13	−0.13	0.13	0.13	0.43	0.43
7.00	0.43	0.43	0.13	0.13	0.56	0.56	0.42	0.42	−0.39	−0.39
8.00	−0.39	−0.39	0.42	0.42	0.04	0.04	0.04	0.04	−0.39	−0.39

* Rounded to two decimal places

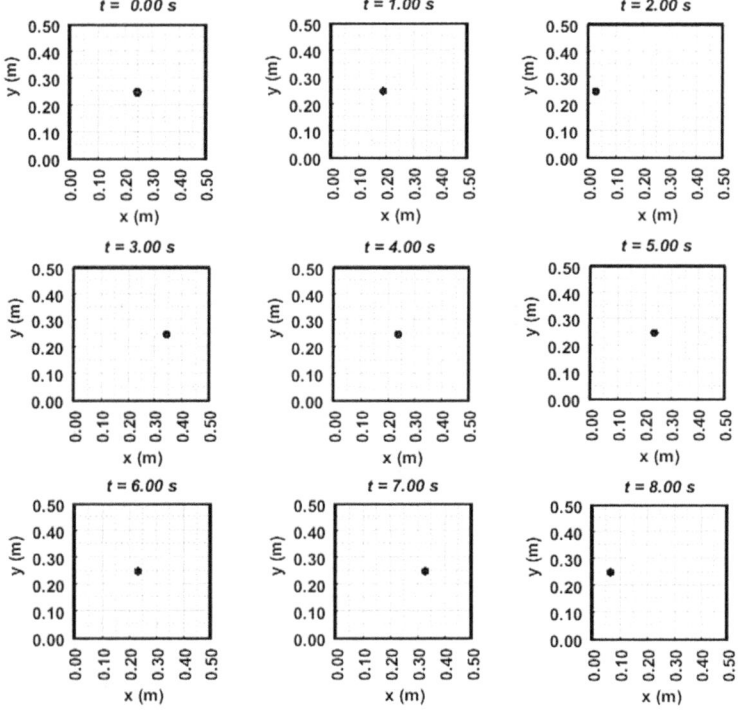

Fig. A.7 Positions of the centre of mass in the first eight seconds of the 6th test

Table A.6 Results of the sixth validation test (SI units)*

t	Input data						Output data			
	V_{ox}	V_{oy}	C_{ox}	C_{oy}	C_{1x}	C_{1y}	C_{fx}	C_{fy}	V_{fx}	V_{fy}
1.00	0.10	0.00	0.25	0.25	0.13	0.25	0.19	0.25	0.81	0.00
2.00	0.81	0.00	0.19	0.25	0.10	0.25	0.03	0.25	−0.73	0.00
3.00	−0.73	0.00	0.03	0.25	−0.70	0.25	0.35	0.25	−0.59	0.00
4.00	−0.59	0.00	0.35	0.25	−0.25	0.25	0.24	0.25	0.53	0.00
5.00	0.53	0.00	0.24	0.25	0.777	0.25	0.24	0.25	−0.48	0.00
6.00	−0.48	0.00	0.24	0.25	−0.24	0.25	0.24	0.25	0.43	0.00
7.00	0.43	0.00	0.24	0.25	0.67	0.25	0.33	0.25	−0.39	0.00
8.00	−0.39	0.00	0.33	0.25	−0.56	0.25	0.07	0.25	0.35	0.00

* Rounded to two decimal places

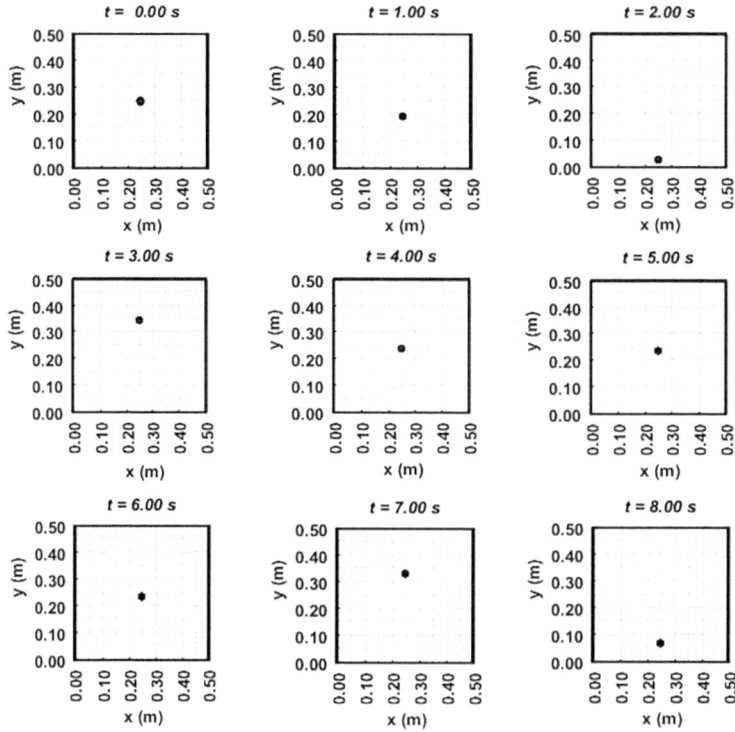

Fig. A.8 Positions of the centre of mass in the first eight seconds of the 7th test

Table A.7 Results of the seventh validation test (SI units)*

t	Input data						Output data			
	V_{ox}	V_{oy}	C_{ox}	C_{oy}	C_{1x}	C_{1y}	C_{fx}	C_{fy}	V_{fx}	V_{fy}
1.00	0.00	1.00	0.25	0.25	0.25	0.13	0.25	0.19	0.00	0.81
2.00	0.00	0.81	0.25	0.19	0.25	1.00	0.25	0.03	0.00	−0.73
3.00	0.00	−0.73	0.25	0.03	0.25	−0.70	0.25	0.35	0.00	−0.59
4.00	0.00	−0.59	0.25	0.35	0.25	−0.25	0.25	0.24	0.00	0.53
5.00	0.00	0.53	0.25	0.24	0.25	0.77	0.25	0.24	0.00	−0.48
6.00	0.00	−0.48	0.25	0.24	0.25	−0.24	0.25	0.24	0.00	0.43
7.00	0.00	0.43	0.25	0.24	0.25	0.67	0.25	0.33	0.00	−0.38
8.00	0.00	−0.38	0.25	0.33	0.25	−0.56	0.25	0.07	0.00	0.35

* Rounded to two decimal places

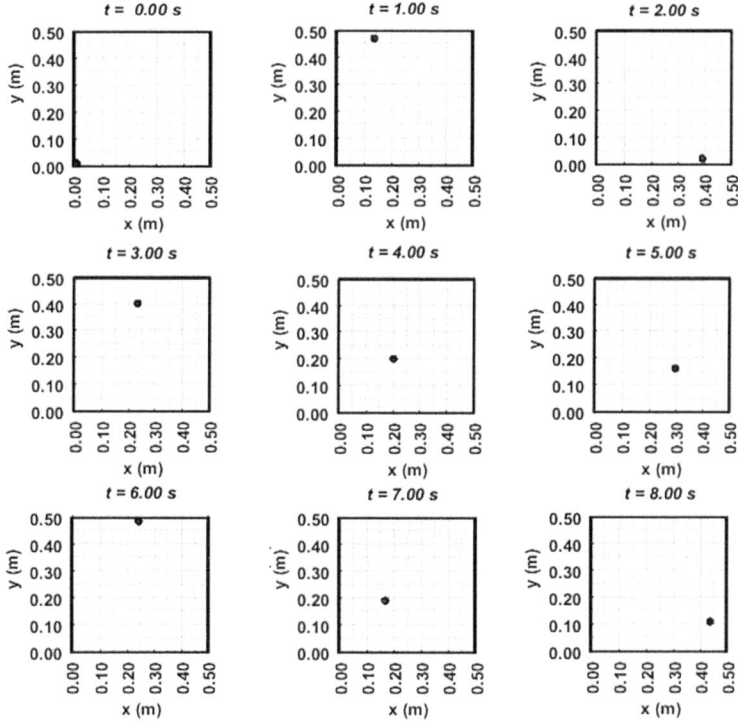

Fig. A.9 Positions of the centre of mass in the first eight seconds of the 8th test

Table A.8 Results of the eighth validation test (SI units)*

t	Input data						Output data			
	V_{ox}	V_{oy}	C_{ox}	C_{oy}	C_{1x}	C_{1y}	C_{fx}	C_{fy}	V_{fx}	V_{fy}
1.00	0.87	0.50	0.01	0.01	0.88	0.51	0.14	0.48	−0.78	−0.45
2.00	−0.78	−0.45	0.14	0.48	−0.64	0.02	0.39	0.02	−0.63	−0.45
3.00	−0.63	−0.45	0.39	0.02	−0.24	−0.42	0.24	0.40	0.57	0.41
4.00	0.57	0.41	0.24	0.40	0.81	0.81	0.20	0.20	−0.51	−0.36
5.00	−0.51	−0.36	0.20	0.20	−0.31	−0.16	0.30	0.16	0.46	0.33
6.00	0.46	0.33	0.30	0.16	0.76	0.49	0.25	0.49	−0.42	−0.30
7.00	−0.42	−0.30	0.25	0.49	−0.17	0.19	0.17	0.19	0.37	−0.30
8.00	0.37	−0.30	0.17	0.19	0.55	−0.10	0.44	0.11	−0.33	0.27

* Rounded to two decimal places

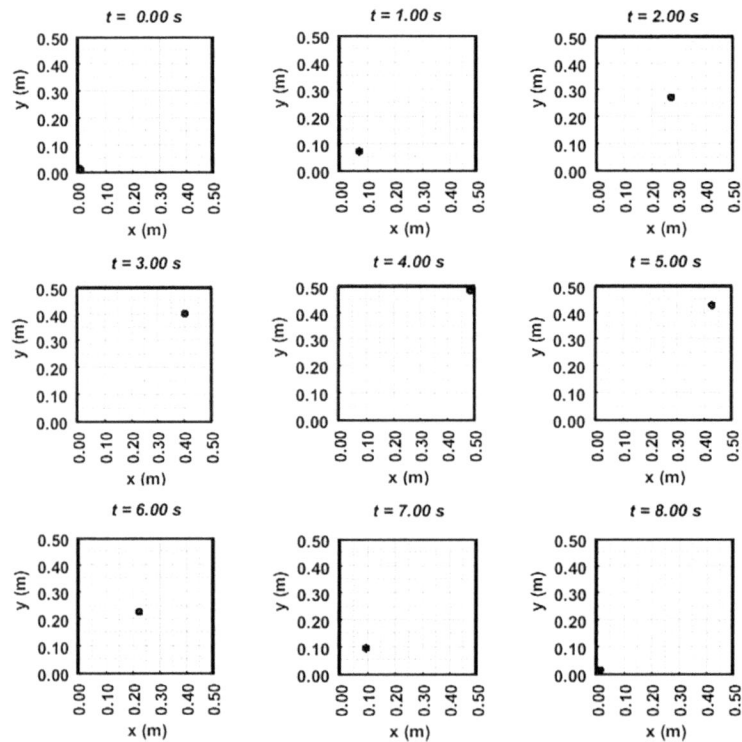

Fig. A.10 Positions of the centre of mass in the first eight seconds of the 9th test

Table A.9 Results of the ninth validation test (SI units)*

t	Input data						Output data			
	V_{ox}	V_{oy}	C_{ox}	C_{oy}	C_{1x}	C_{1y}	C_{fx}	C_{fy}	V_{fx}	V_{fy}
1.00	1.00	1.00	0.01	0.01	1.01	1.01	0.07	0.07	−1.00	−1.00
2.00	−1.00	−1.00	0.07	0.07	−0.93	−0.93	0.28	0.28	−1.00	−1.00
3.00	−1.00	−1.00	0.28	0.28	−0.73	−0.73	0.40	0.40	−1.00	−1.00
4.00	−1.00	−1.00	0.40	0.40	−0.59	−0.59	0.49	0.49	−1.00	−1.00
5.00	−1.00	−1.00	0.49	0.49	−0.51	−0.51	0.43	0.43	1.00	1.00
6.00	1.00	1.00	0.43	0.43	1.43	1.43	0.23	0.23	1.00	1.00
7.00	1.00	1.00	0.23	0.23	0.12	0.12	0.10	0.10	1.00	1.00
8.00	1.00	1.00	0.10	0.10	1.10	1.10	0.02	0.02	1.00	1.00

* Rounded to two decimal places

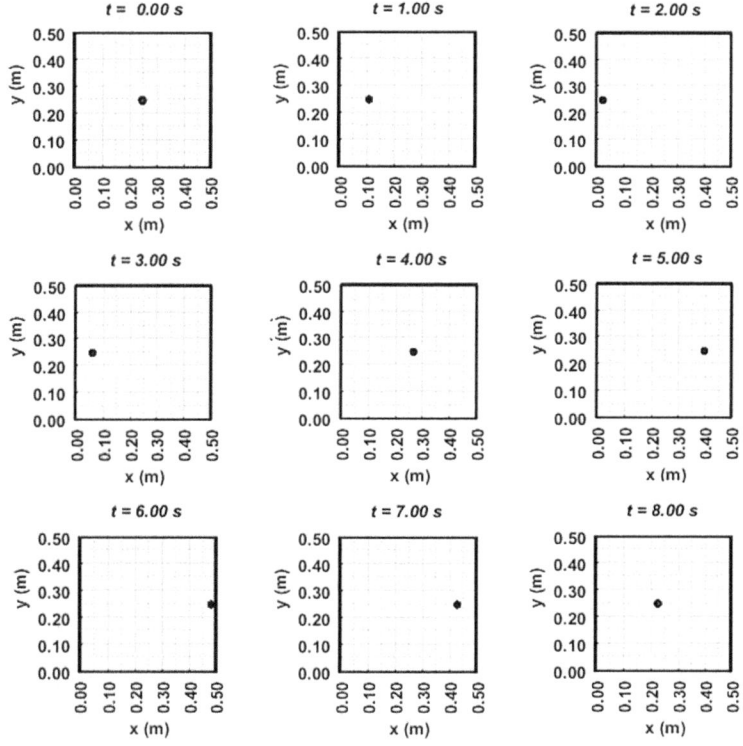

Fig. A.11 Positions of the centre of mass in the first eight seconds of the 10th test

Table A.10 Results of the 10th validation test (SI units)*

t	Input data						Output data			
	V_{ox}	V_{oy}	C_{ox}	C_{oy}	C_{1x}	C_{1y}	C_{fx}	C_{fy}	V_{fx}	V_{fy}
1.00	1.00	0.00	0.25	0.25	1.25	0.25	0.11	0.25	1.00	0.00
2.00	1.00	0.00	0.11	0.25	1.11	0.25	0.02	0.25	1.00	0.00
3.00	1.00	0.00	0.02	0.25	1.02	0.25	0.06	0.25	−1.00	0.00
4.00	−1.00	0.00	0.06	0.25	−0.94	0.25	0.27	0.25	−1.00	0.00
5.00	−1.00	0.00	0.27	0.25	−0.73	0.25	0.40	0.25	−1.00	0.00
6.00	−1.00	0.00	0.40	0.25	−0.60	0.25	0.48	0.25	−1.00	0.00
7.00	−1.00	0.00	0.48	0.25	−0.52	0.25	0.43	0.25	1.00	0.00
8.00	1.00	0.00	0.43	0.25	1.43	0.25	0.23	0.25	1.00	0.00

* Rounded to two decimal places

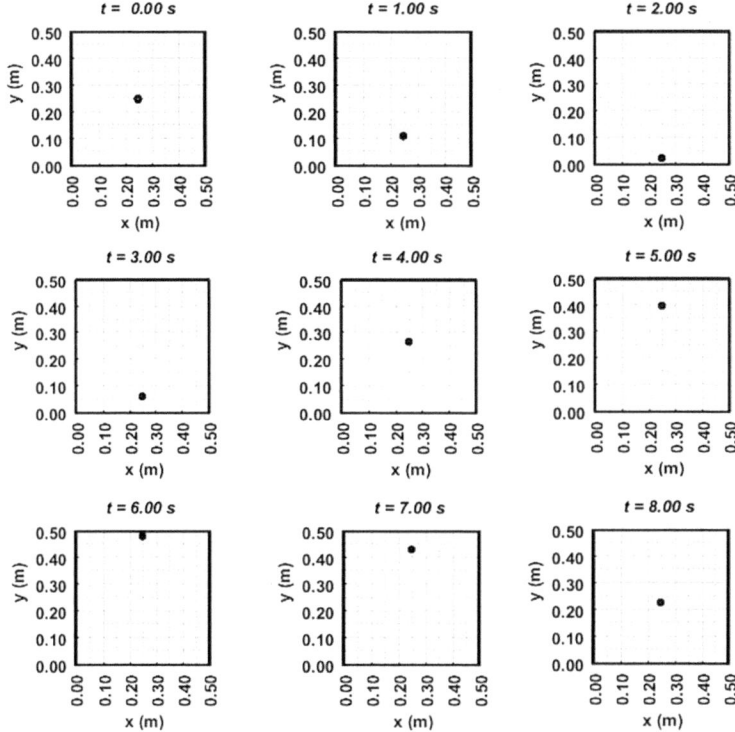

Fig. A.12 Positions of the centre of mass in the first eight seconds of the 11th test

Table A.11 Results of the 11th validation test (SI units)*

t	Input data						Output data			
	V_{ox}	V_{oy}	C_{ox}	C_{oy}	C_{1x}	C_{1y}	C_{fx}	C_{fy}	V_{fx}	V_{fy}
1.00	0.00	1.00	0.25	0.25	0.25	1.25	0.25	0.11	0.00	1.00
2.00	0.00	1.00	0.25	0.11	0.25	1.11	0.25	0.02	0.00	1.00
3.00	0.00	1.00	0.25	0.02	0.25	1.02	0.25	0.06	0.00	−1.00
4.00	0.00	−1.00	0.25	0.06	0.25	0.77	0.25	0.27	0.00	−1.00
5.00	0.00	−1.00	0.25	0.27	0.25	−0.73	0.25	0.40	0.00	−1.00
6.00	0.00	−1.00	0.25	0.40	0.25	−0.60	0.25	0.48	0.00	−1.00
7.00	0.00	−1.00	0.25	0.48	0.25	−0.52	0.25	0.43	0.00	1.00
8.00	0.00	1.00	0.25	0.43	0.25	1.43	0.25	0.23	0.00	1.00

* Rounded to two decimal places

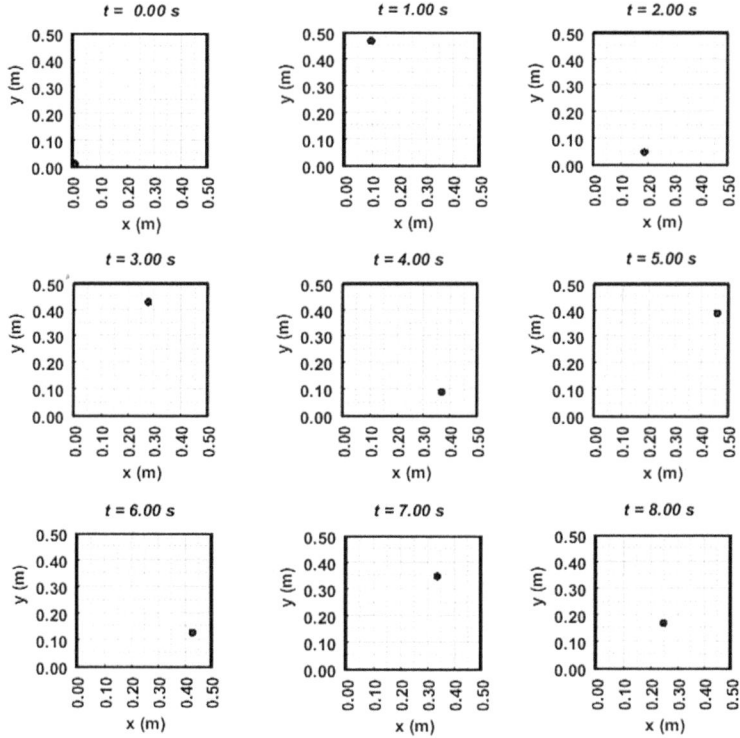

Fig. A.13 Positions of the centre of mass in the first eight seconds of the 12th test

Table A.12 Results of the 12th validation test (SI units)*

t	Input data						Output data			
	V_{ox}	V_{oy}	C_{ox}	C_{oy}	C_{1x}	C_{1y}	C_{fx}	C_{fy}	V_{fx}	V_{fy}
1.00	0.87	0.50	0.01	0.01	0.88	0.51	0.10	0.47	−0.87	−0.50
2.00	v0.87	−0.50	0.10	0.47	−0.77	−0.30	0.19	0.05	−0.87	0.50
3.00	−0.87	0.50	0.19	0.05	−0.68	0.55	0.28	0.43	−0.87	−0.50
4.00	−0.87	−0.50	0.28	0.43	−0.59	−0.07	0.37	0.09	−0.87	0.50
5.00	−0.87	0.50	0.37	0.09	−0.50	0.59	0.46	0.39	−0.87	−0.50
6.00	−0.87	−0.50	0.46	0.39	−0.41	−0.11	0.43	0.13	0.87	0.50
7.00	0.87	0.50	0.43	0.13	0.13	0.63	0.34	0.35	0.87	−0.50
8.00	0.87	−0.50	0.34	0.35	1.21	−0.15	0.25	0.17	0.87	0.50

* Rounded to two decimal places

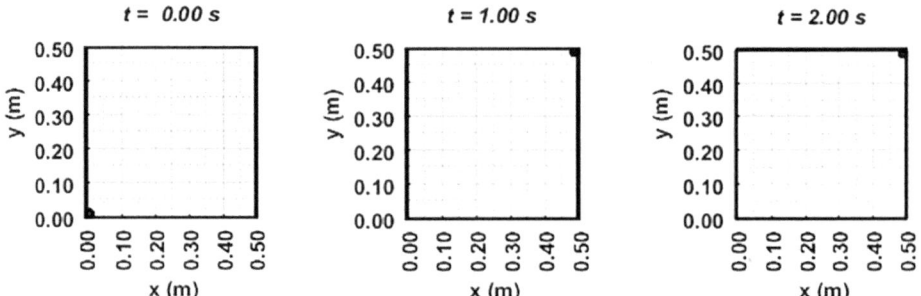

Fig. A.14 Positions of the centre of mass in the first two seconds of the 13th test

Test 13:

Initial position of the centre of mass: $C_o = (0.01, 0.01)$ m.

Initial velocity: $V_o = (0.10, 0.10)$ m/s.

CR = 0.00 (Fig. A.14 and Table A.13).

Test 14:

Initial position of the centre of mass: $C_o = (0.01, 0.01)$ m.

Initial velocity: $V_o = (0.10, 0.00)$ m/s.

CR = 0.00 (Fig. A.15 and Table A.14).

Table A.13 Results of the 13th validation test (SI units)*

t	Input data						Output data			
	V_{ox}	V_{oy}	C_{ox}	C_{oy}	C_{1x}	C_{1y}	C_{fx}	C_{fy}	V_{fx}	V_{fy}
1.00	1.00	1.00	0.01	0.01	1.01	1.01	0.49	0.49	0.00	0.00
2.00	0.00	0.00	0.49	0.49	0.49	0.49	0.49	0.49	0.00	0.00
3.00	0.00	0.00	0.49	0.49	0.49	0.49	0.49	0.49	0.00	0.00

* Rounded to two decimal places

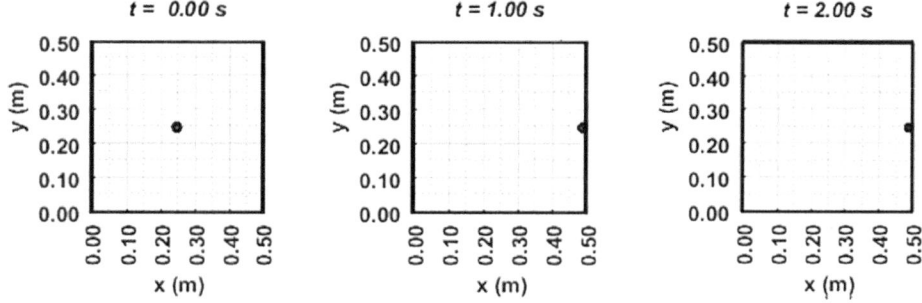

Fig. A.15 Positions of the centre of mass in the first two seconds of the 14th test

Table A.14 Results of the 14th validation test (SI units)*

t	Input data						Output data			
	V_{ox}	V_{oy}	C_{ox}	C_{oy}	C_{1x}	C_{1y}	C_{fx}	C_{fy}	V_{fx}	V_{fy}
1.00	1.00	0.00	0.25	0.25	1.25	0.25	0.49	0.25	0.00	0.00
2.00	0.00	0.00	0.49	0.25	0.49	0.25	0.49	0.25	0.00	0.00
3.00	0.00	0.00	0.49	0.25	0.49	0.25	0.49	0.25	0.00	0.00

* Rounded to two decimal places

A wide range of tests were performed, and some of these results were presented here. The output provided by the algorithm (positions of the centre of mass and particle velocities at each numerical iteration) has been verified using analytical results; the numerical and analytical results were in agreement.

Appendix B
CDRA Validation in Three-Dimensional Domains

Tests were performed, and the particle had a radius of 0.01 m. The three-dimensional closed box had sides of 0.50 m, as shown in Fig. B.1. The time step used in tests was 0.25 s. Different values were assigned to the coefficients of restitution of kinetic energy (CR) and friction (CF), and test results are presented below.

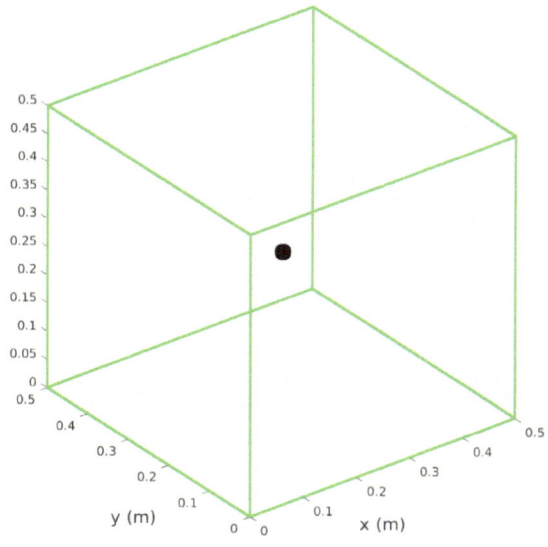

Fig. B.1 The closed box used in the algorithm validation tests and the initial position of a particle. Reproduced from Fraga Filho, C. A. D, et al. Implementation of three-dimensional physical reflective boundary conditions in mesh-free particle methods for continuum fluid dynamics: Validation tests and case studies. Phys. Fluids 31, 103,606 (2019), with the permission of AIP

C. A. D. Fraga Filho, *Reflective Boundary Conditions in SPH Fluid Dynamics Simulation*, Synthesis Lectures on Mechanical Engineering, https://doi.org/10.1007/978-3-031-71582-2

Fig. B.2 Positions of the centre of mass of the particle at the 1st validation test

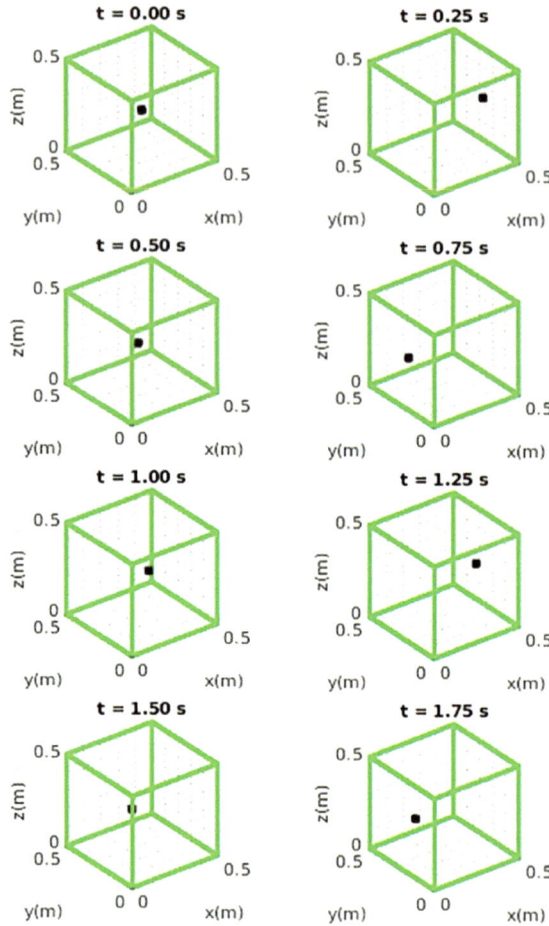

Test 1:

Initial position of the centre of mass: $C_o = (0.25, 0.25, 0.25)$ m; Initial velocity: $V_o = (1.00, 0.00, 0.00)$ m/s; CR $= 1.00$; CF $= 0.00$ (the simulation results are in Table B.1 and Fig. B.2).

Test 2:

Initial position of the centre of mass: $C_o = (0.25, 0.25, 0.25)$ m; Initial velocity: $V_o = (0.00, 1.00, 0.00)$ m/s; CR $= 1.00$; CF $= 0.00$ the simulation results are in Table B.2 and Fig. B.3).

Table B.1 Results of the 1st validation test* (SI units)

t **	Input data									Output data					
	V_{ox}	V_{oy}	V_{oz}	C_{ox}	C_{oy}	C_{oz}	C_{1x}	C_{1y}	C_{1z}	C_{fx}	C_{fy}	C_{fz}	V_{fx}	V_{fy}	V_{fz}
0.25	1.00	0.00	0.00	0.25	0.25	0.25	0.50	0.25	0.25	0.48	0.25	0.25	−1.00	0.00	0.00
0.50	−1.00	0.00	0.00	0.48	0.25	0.25	0.23	0.25	0.25	0.23	0.25	0.25	−1.00	0.00	0.00
0.75	−1.00	0.00	0.00	0.23	0.25	0.25	−0.02	0.25	0.25	0.04	0.25	0.25	1.00	0.00	0.00
1.00	1.00	0.00	0.00	0.04	0.25	0.25	0.29	0.25	0.25	0.29	0.25	0.25	1.00	0.00	0.00
1.25	1.00	0.00	0.00	0.29	0.25	0.25	0.54	0.25	0.25	0.44	0.25	0.25	−1.00	0.00	0.00
1.50	−1.00	0.00	0.00	0.44	0.25	0.25	0.19	0.25	0.25	0.19	0.25	0.25	−1.00	0.00	0.00
1.75	−1.00	0.00	0.00	0.19	0.25	0.25	−0.06	0.25	0.25	0.08	0.25	0.25	1.00	0.00	0.00
2.00	1.00	0.00	0.00	0.08	0.25	0.25	0.33	0.25	0.25	0.33	0.25	0.25	1.00	0.00	0.00

* Rounded to two decimal places
** Number of collisions in every timestep: 1st—01; 2nd—0; 3rd—01; 4th—0; 5th—01; 6th—0; 7th—01; 8th—0

Fig. B.3 Positions of the centre of mass of the particle at the 2nd validation test

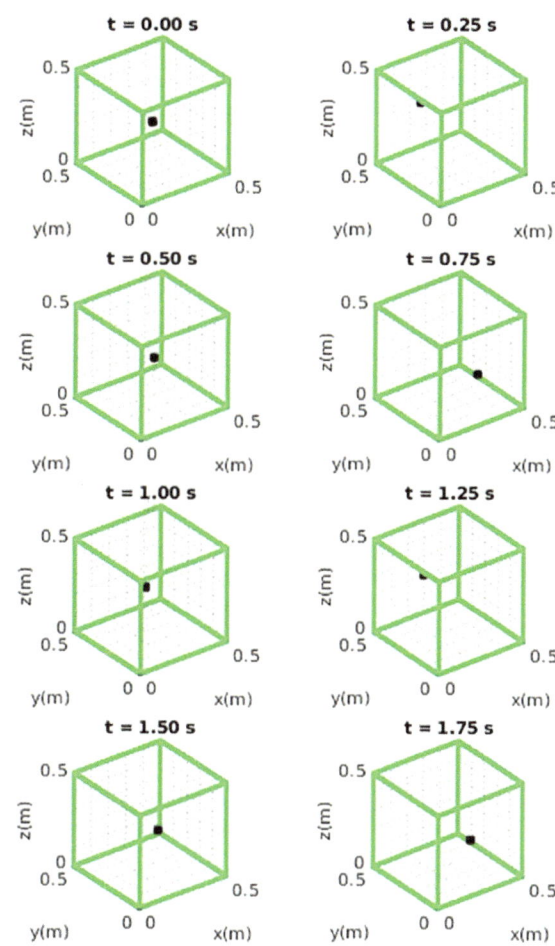

Table B.2 Results of the 2nd validation test* (SI units)

t **	Input data									Output data					
	V_{ox}	V_{oy}	V_{oz}	C_{ox}	C_{oy}	C_{oz}	C_{1x}	C_{1y}	C_{1z}	C_{fx}	C_{fy}	C_{fz}	V_{fx}	V_{fy}	V_{fz}
0.25	0.00	1.00	0.00	0.25	0.25	0.25	0.25	0.50	0.25	0.25	0.48	0.25	0.00	−1.00	0.00
0.50	0.00	−1.00	0.00	0.25	0.48	0.25	0.25	0.23	0.25	0.25	0.23	0.25	0.00	−1.00	0.00
0.75	0.00	−1.00	0.00	0.25	0.23	0.25	0.25	−0.02	0.25	0.25	0.04	0.25	0.00	1.00	0.00
1.00	0.00	1.00	0.00	0.25	0.04	0.25	0.25	0.29	0.25	0.25	0.29	0.25	0.00	1.00	0.00
1.25	0.00	1.00	0.00	0.25	0.29	0.25	0.25	0.54	0.25	0.25	0.44	0.25	0.00	−1.00	0.00
1.50	0.00	−1.00	0.00	0.25	0.44	0.25	0.25	0.19	0.25	0.25	0.19	0.25	0.00	−1.00	0.00
1.75	0.00	−1.00	0.00	0.25	0.19	0.25	0.25	−0.06	0.25	0.25	0.08	0.25	0.00	1.00	0.00
2.00	0.00	1.00	0.00	0.25	0.08	0.25	0.25	0.33	0.25	0.25	0.33	0.25	0.00	1.00	0.00

* Rounded to two decimal places
** Number of collisions in every timestep: 1st—01; 2nd—0; 3rd—01; 4th—0; 5th—01; 6th—0; 7th—01; 8th—0

Test 3:

Initial position of the centre of mass: C_o = (0.25, 0.25, 0.25) m; Initial velocity: V_o = (0.00, 0.00, 1.00) m/s; CR = 1.00; CF = 0.00 (the simulation results are in Table B.3 and Fig. B.4).

Test 4:

Initial position of the centre of mass: C_o = (0.01, 0.01, 0.01) m; Initial velocity: V_o = (1.00, 1.00, 1.00) m/s; CR = 1.00; CF = 0.00 (the simulation results are in Table B.4 and Fig. B.5).

Test 5:

Initial position of the centre of mass: C_o = (0.01, 0.01, 0.01) m; Initial velocity: V_o = (1.00, 1.00, 0.00) m/s; CR = 1.00; CF = 0.00 (the simulation results are in Table B.5 and Fig. B.6).

Test 6:

Initial position of the centre of mass: C_o = (0.49, 0.49, 0.49) m; Initial velocity: V_o = (−1.00, −1.00, −1.00) m/s; CR = 1.00; CF = 0.00 (the simulation results are in Table B.6 and Fig. B.7).

Test 7:

Initial position of the centre of mass: C_o = (0.01, 0.01, 0.49) m; Initial velocity: V_o = (1.00, 1.00, −1.00) m/s; CR = 1.00; CF = 0.00 (the simulation results are in Table B.7 and Fig. B.8).

Fig. B.4 Positions of the centre of mass of the particle at the 3rd validation test

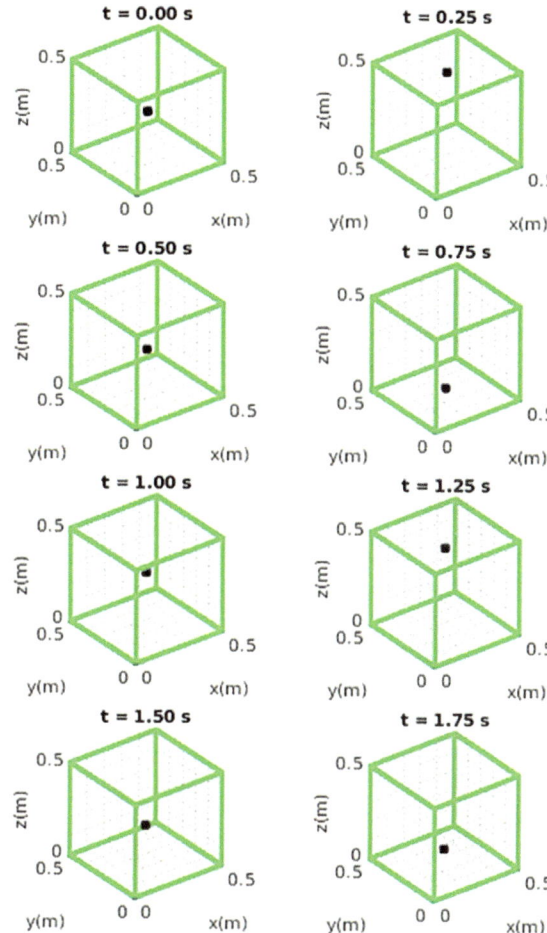

Table B.3 Results of the 3rd validation test* (SI units)

t **	Input data									Output data					
	V_{ox}	V_{oy}	V_{oz}	C_{ox}	C_{oy}	C_{oz}	C_{1x}	C_{1y}	C_{1z}	C_{fx}	C_{fy}	C_{fz}	V_{fx}	V_{fy}	V_{fz}
0.25	0.00	0.00	1.00	0.25	0.25	0.25	0,25	0.25	0.50	0.25	0.25	0.48	0.00	0.00	−1.00
0.50	0.00	0.00	−1.00	0.25	0.25	0.48	0.25	0.25	0.23	0.25	0.25	0.23	0.00	0.00	−1.00
0.75	0.00	0.00	−1.00	0.25	0.25	0.23	0.25	0.25	−0.02	0.25	0.25	0.04	0.00	0.00	1.00
1.00	0.00	0.00	1.00	0.25	0.25	0.04	0.25	0.25	0.29	0.25	0.25	0.29	0.00	0.00	1.00
1.25	0.00	0.00	1.00	0.25	0.25	0.29	0.25	0.25	0.54	0.25	0.25	0.44	0.00	0.00	−1.00
1.50	0.00	0.00	−1.00	0.25	0.25	0.44	0.25	0.25	0.19	0.25	0.25	0.19	0.00	0.00	−1.00
1.75	0.00	0.00	−1.00	0.25	0.25	0.19	0.25	0.25	−0.06	0.25	0.25	0.08	0.00	0.00	1.00
2.00	0.00	0.00	1.00	0.25	0.25	0.08	0.25	0.25	0.33	0.25	0.25	0.33	0.00	0.00	1.00

* Rounded to two decimal places
** Number of collisions in every timestep: 1st—01; 2nd—0; 3rd—01; 4th—0; 5th—01; 6th—0; 7th—01; 8th—0

Fig. B.5 Positions of the centre of mass of the particle at the 4th validation test

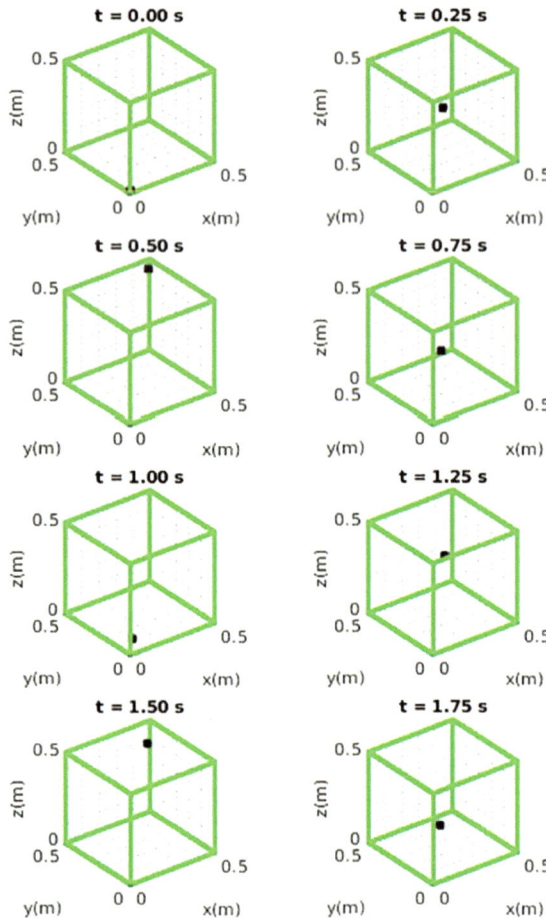

Test 8:

Initial position of the centre of mass: $C_o = (0.01, 0.01, 0.01)$ m; Initial velocity: $V_o = (1.50, 1.00, 1.00)$ m/s; CR = 1.00; CF = 0.10 (the simulation results are in Table B.8 and Fig. B.9).

Test 9:

Initial position of the centre of mass: $C_o = (0.25, 0.25, 0.25)$ m; Initial velocity: $V_o = (-1.50, -1.50, 1.00)$ m/s; CR = 1.00; CF = 0.10 (the simulation results are in Table B.9 and Fig. B.10).

Table B.4 Results of the 4th validation test* (SI units)

t **	Input data									Output data					
	V_{ox}	V_{oy}	V_{oz}	C_{ox}	C_{oy}	C_{oz}	C_{1x}	C_{1y}	C_{1z}	C_{fx}	C_{fy}	C_{fz}	V_{fx}	V_{fy}	V_{fz}
0.25	1.00	1.00	1.00	0.01	0.01	0.01	0.26	0.26	0.26	0.26	0.26	0.26	1.00	1.00	1.00
0.50	1.00	1.00	1.00	0.26	0.26	0.26	0.51	0.51	0.51	0.47	0.47	0.47	-1.00	-1.00	-1.00
0.75	-1.00	-1.00	-1.00	0.47	0.47	0.47	0.22	0.22	0.22	0.22	0.22	0.22	-1.00	-1.00	-1.00
1.00	-1.00	-1.00	-1.00	0.22	0.22	0.22	-0.03	-0.03	-0.03	0.05	0.05	0.05	1.00	1.00	1.00
1.25	1.00	1.00	1.00	0.05	0.05	0.05	0.30	0.30	0.30	0.30	0.30	0.30	1.00	1.00	1.00
1.50	1.00	1.00	1.00	0.30	0.30	0.30	0.55	0.55	0.55	0.43	0.43	0.43	-1.00	-1.00	-1.00
1.75	-1.00	-1.00	-1.00	0.43	0.43	0.43	0.18	0.18	0.18	0.18	0.18	0.18	-1.00	-1.00	-1.00
2.00	-1.00	-1.00	-1.00	0.18	0.18	0.18	-0.07	-0.07	-0.07	0.09	0.09	0.09	1.00	1.00	1.00

* Rounded to two decimal places

** Number of collisions in every timestep: 1st—0; 2nd—01; 3rd—0; 4th—01; 5th—0; 6th—01; 7th—0; 8th—01

Fig. B.6 Positions of the centre of mass of the particle at the 5th validation test

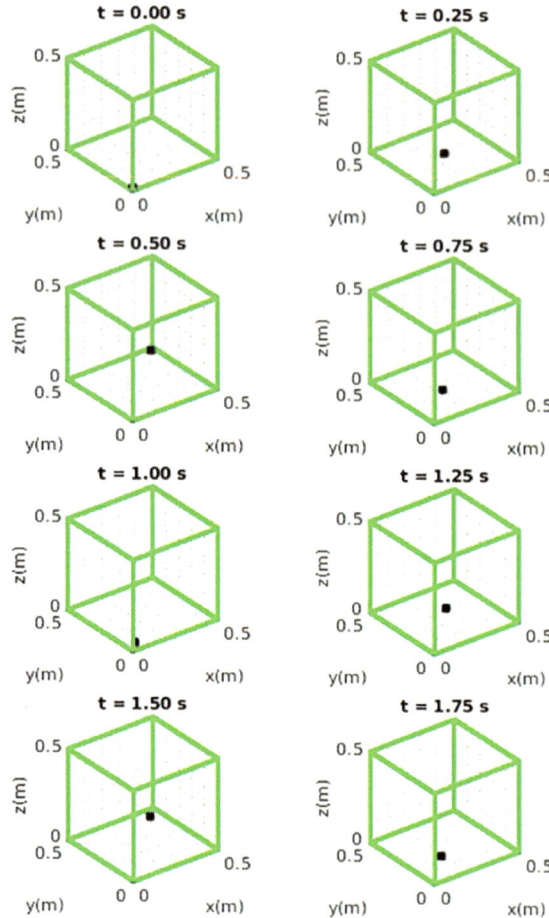

Test 10:

Initial position of the centre of mass: $C_o = (0.25, 0.25, 0.25)$ m; Initial velocity: $V_o = (-1.00, -1.50, -1.50)$ m/s; CR $= 0.90$; CF $= 0.10$ (the simulation results are in Table B.10 and Fig. B.11).

Test 11:

Initial position of the centre of mass: $C_o = (0.25, 0.25, 0.25)$ m; Initial velocity: $V_o = (1.50, 1.50, 1.00)$ m/s; CR $= 0.80$; CF $= 0.10$ (the simulation results are in Table B.11 and Fig. B.12).

Table B.5 Results of the 5th validation test* (SI units)

t **	Input data									Output data					
	V_{ox}	V_{oy}	V_{oz}	C_{ox}	C_{oy}	C_{oz}	C_{1x}	C_{1y}	C_{1z}	C_{fx}	C_{fy}	C_{fz}	V_{fx}	V_{fy}	V_{fz}
0.25	1.00	1.00	0.00	0.01	0.01	0.01	0.26	0.26	0.01	0.26	0.26	0.01	1.001	1.00	0.00
0.50	1.001	1.00	0.00	0.26	0.26	0.01	0.51	0.51	0.01	0.47	0.47	0.01	−1.00	−1.00	0.00
0.75	−1.00	−1.00	0.00	0.47	0.47	0.01	0.22	0.22	0.01	0.22	0.22	0.01	−1.00	−1.00	0.00
1.00	−1.00	−1.00	0.00	0.22	0.22	0.01	−0.03	−0.03	0.01	0.05	0.05	0.01	1.00	1.00	0.00
1.25	1.00	1.00	0.00	0.05	0.05	0.01	0.30	0.30	0.01	0.30	0.30	0.01	1.00	1.00	0.00
1.50	1.00	1.00	0.00	0.30	0.30	0.01	0.55	0.55	0.01	0.43	0.43	0.01	−1.00	−1.00	0.00
1.75	−1.00	−1.00	0.00	0.43	0.43	0.01	0.18	0.18	0.01	0.18	0.18	0.01	−1.00	−1.00	0.00
2.00	−1.00	−1.00	0.00	0.18	0.18	0.01	−0.07	−0.07	0.01	0.09	0.09	0.01	1.00	1.00	0.00

* Rounded to two decimal places

** Number of collisions in every timestep: 1st—0; 2nd—01; 3rd—0; 4th—01; 5th—0; 6th—01; 7th—0; 8th—01

Fig. B.7 Positions of the centre of mass of the particle at the 6th validation test

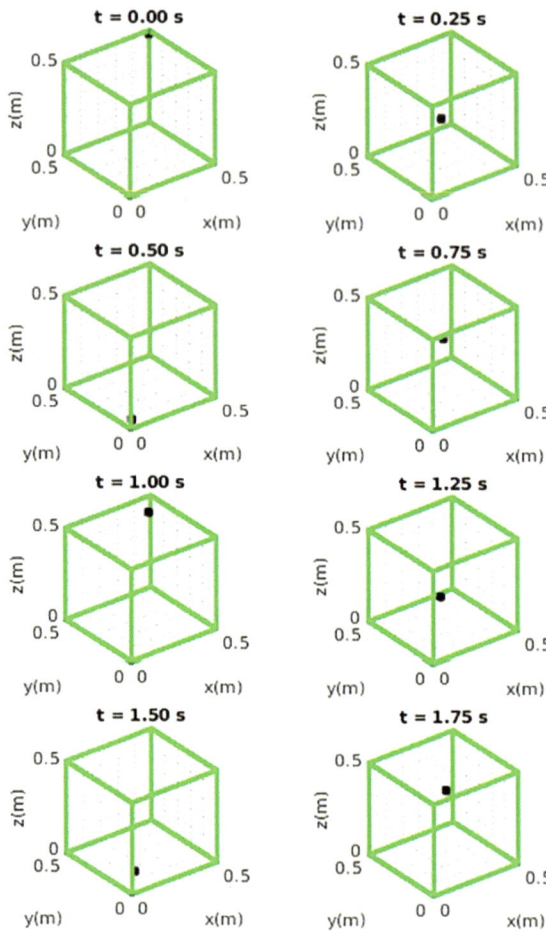

Test 12:

Initial position of the centre of mass: $C_o = (0.01, 0.25, 0.25)$ m; Initial velocity: $V_o = (1.50, 1.50, 0.00)$ m/s; CR = 0.80; CF = 0.20 (the simulation results are in Table B.12 and Fig. B.13).

Test 13:

Initial position of the centre of mass: $C_o = (0.01, 0.01, 0.25)$ m; Initial velocity: $V_o = (1.50, 1.50, 1.50)$ m/s; CR = 0.80; CF = 0.20 (the simulation results are in Table B.13 and Fig. B.14).

Table B.6 Results of the 6th validation test* (SI units)

t**	Input data									Output data					
	V_{ox}	V_{oy}	V_{oz}	C_{ox}	C_{oy}	C_{oz}	C_{1x}	C_{1y}	C_{1z}	C_{fx}	C_{fy}	C_{fz}	V_{fx}	V_{fy}	V_{fz}
0.25	−1.00	−1.00	−1.00	0.49	0.49	0.49	0.24	0.24	0.24	0.24	0.24	0.24	−1.00	−1.00	−1.00
0.50	−1.00	−1.00	−1.00	0.24	0.24	0.24	−0.01	−0.01	−0.01	0.03	0.03	0.03	1.00	1.00	1.00
0.75	1.00	1.00	1.00	0.03	0.03	0.03	0.28	0.28	0.28	0.28	0.28	0.28	1.00	1.00	1.00
1.00	1.00	1.00	1.00	0.28	0.28	0.28	0.53	0.53	0.53	0.45	0.45	0.45	−1.00	−1.00	−1.00
1.25	−1.00	−1.00	−1.00	0.45	0.45	0.45	0.20	0.20	0.20	0.20	0.20	0.20	−1.00	−1.00	−1.00
1.50	−1.00	−1.00	−1.00	0.20	0.20	0.20	−0.05	−0.05	−0.05	0.07	0.07	0.07	1.00	1.00	1.00
1.75	1.00	1.00	1.00	0.07	0.07	0.07	0.32	0.32	0.32	0.32	0.32	0.32	1.00	1.00	1.00
2.00	1.00	1.00	1.00	0.32	0.32	0.32	0.57	0.57	0.57	0.41	0.41	0.41	−1.00	−1.00	−1.00

* Rounded to two decimal places
** Number of collisions in every timestep: 1st—0; 2nd—01; 3rd—0; 4th—01; 5th—0; 6th—01; 7th—0; 8th—01

Fig. B.8 Positions of the
centre of mass of the particle at
the 7th validation test

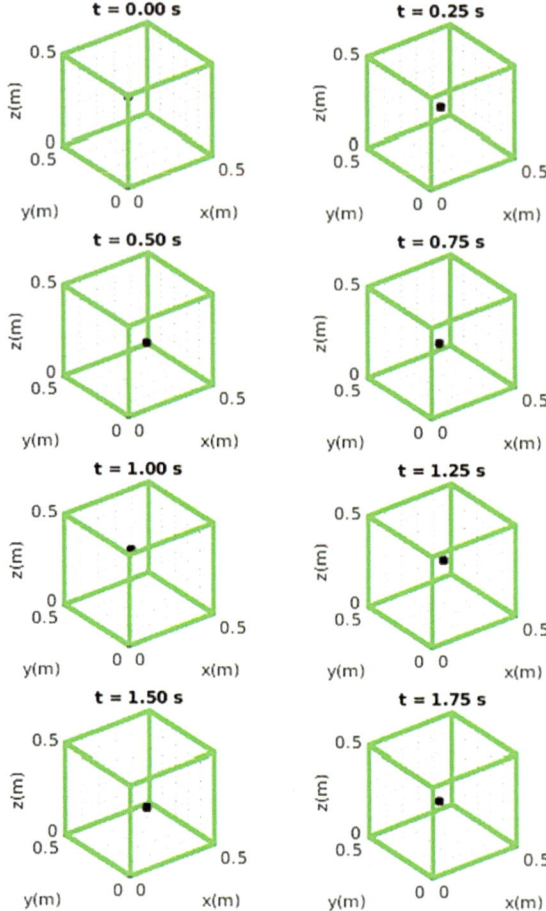

Test 14:

Initial position of the centre of mass: $C_o = (0.25, 0.25, 0.49)$ m; Initial velocity: $V_o = (-1.50, -1.50, -1.50)$ m/s; (the simulation results are in Table B.14 and Fig. B.15).

Test 15:

Initial position of the centre of mass: $C_o = (0.01, 0.01, 0.49)$ m; Initial velocity: $V_o = (1.50, 1.50, 1.00)$ m/s; $CR = 0.85$; $CF = 0.15$ (the simulation results are in Table B.15 and Fig. B.16).

Table B.7 Results of the 7th validation test* (SI units)

t**	Input data									Output data					
	V_{ox}	V_{oy}	V_{oz}	C_{ox}	C_{oy}	C_{oz}	C_{1x}	C_{1y}	C_{1z}	C_{fx}	C_{fy}	C_{fz}	V_{fx}	V_{fy}	V_{fz}
0.25	1.00	1.00	−1.00	0.01	0.01	0.49	0.26	0.26	0.24	0.26	0.26	0.24	1.00	1.00	−1.00
0.50	1.00	1.00	−1.00	0.26	0.26	0.24	0.51	0.51	−0.01	0.47	0.47	0.03	−1.00	−1.00	−1.00
0.75	−1.00	−1.00	−1.00	0.47	0.47	0.03	0.22	0.22	−0.22	0.22	0.22	0.24	−1.00	−1.00	1.00
1.00	−1.00	−1.00	1.00	0.22	0.22	0.24	−0.03	−0.03	0.49	0.05	0.05	0.49	1.00	1.00	1.00
1.25	1.00	1.00	1.00	0.05	0.05	0.49	0.30	0.30	0.74	0.30	0.30	0.24	1.00	1.00	−1.00
1.50	1.00	1.00	−1.00	0.30	0.30	0.24	0.55	0.55	−0.01	0.43	0.43	0.03	−1.00	−1.00	1.00
1.75	−1.00	−1.00	1.00	0.43	0.43	0.03	0.18	0.18	0.28	0.18	0.18	0.28	−1.00	−1.00	1.00
2.00	−1.00	−1.00	1.00	0.18	0.18	0.28	−0.07	−0.07	0.53	0.09	0.09	0.45	1.00	1.00	−1.00

* Rounded to two decimal places
** Number of collisions in every timestep: 1st—0; 2nd—01; 3rd—01; 4th—01; 5th—01; 6th—02; 7th—0; 8th—02

Fig. B.9 Positions of the
centre of mass of the particle at
the 8th validation test

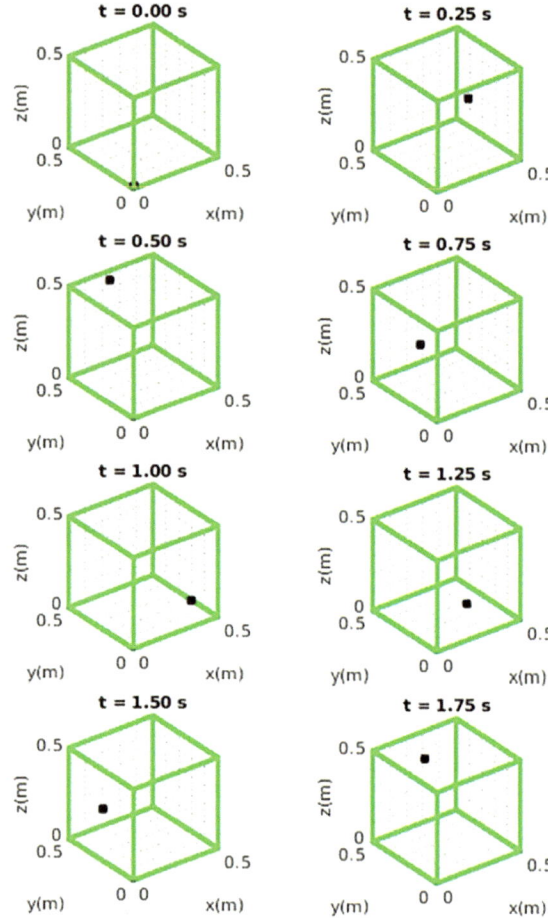

Test 16:

Initial position of the centre of mass: $C_o = (0.49, 0.49, 0.01)$ m; Initial velocity: $V_o = (1.50, 1.50, 1.50)$ m/s; CR $= 0.85$; CF $= 0.15$ (the simulation results are in Table B.16 and Fig. B.17).

Table B.8 Results of the 8th validation test* (SI units)

t **	Input data									Output data					
	V_{ox}	V_{oy}	V_{oz}	C_{ox}	C_{oy}	C_{oz}	C_{1x}	C_{1y}	C_{1z}	C_{fx}	C_{fy}	C_{fz}	V_{fx}	V_{fy}	V_{fz}
0.25	1.50	1.00	1.00	0.01	0.01	0.01	0.39	0.26	0.26	0.39	0.26	0.26	1.50	1.00	1.00
0.50	1.50	1.00	1.00	0.39	0.26	0.26	0.76	0.51	0.51	0.22	0.47	0.47	-1.22	-0.81	-0.81
0.75	-1.22	-0.81	-0.81	0.22	0.47	0.47	-0.08	0.27	0.27	0.10	0.27	0.27	1.22	-0.73	-0.73
1.00	1.22	-0.73	-0.73	0.10	0.27	0.27	0.41	0.09	0.09	0.41	0.09	0.09	1.22	-0.73	v0.73
1.25	1.22	-0.73	-0.73	0.41	0.09	0.09	0.71	-0.10	-0.10	0.27	0.12	0.12	-0.98	0.59	0.59
1.50	-0.98	0.59	0.59	0.27	0.12	0.12	0.02	0.26	0.26	0.02	0.26	0.26	-0.98	0.59	0.59
1.75	-0.98	0.59	0.59	0.02	0.26	0.26	-0.22	0.41	0.41	0.24	0.41	0.41	0.98	0.53	0.53
2.00	0.98	0.53	0.53	0.24	0.41	0.41	0.49	0.55	0.55	0.49	0.43	0.43	0.80	-0.48	-0.48

* Rounded to two decimal places
** Number of collisions in every timestep: 1st—0; 2nd—02; 3rd—01; 4th—0; 5th—02; 6th—0; 7th—01; 8th—01

Fig. B.10 Positions of the centre of mass of the particle at the 9th validation test

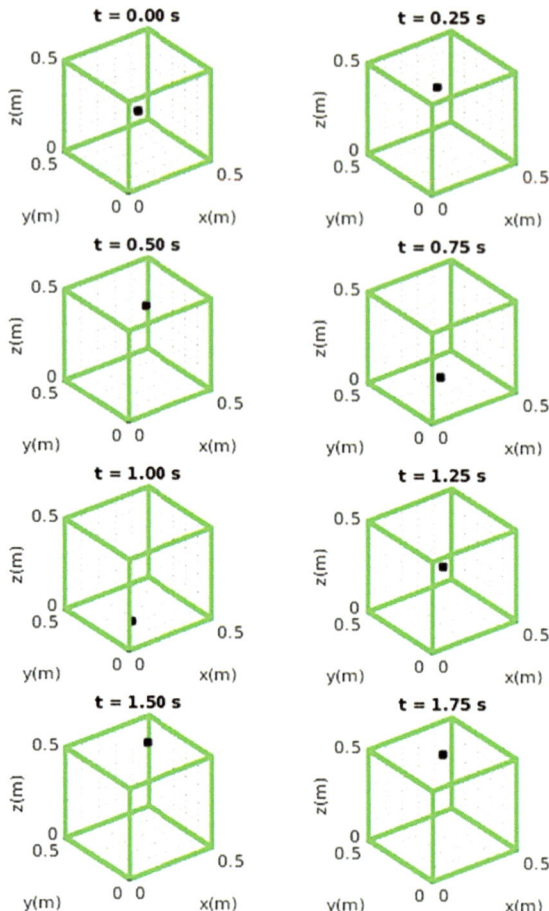

Table B.9 Results of the 9th validation test* (SI units)

t **	Input data									Output data					
	V_{ox}	V_{oy}	V_{oz}	C_{ox}	C_{oy}	C_{oz}	C_{1x}	C_{1y}	C_{1z}	C_{fx}	C_{fy}	C_{fz}	V_{fx}	V_{fy}	V_{fz}
0.25	−1.50	−1.50	1.00	0.25	0.25	0.25	−0.12	−0.12	0.50	0.15	0.15	0.48	1.22	1.22	−0.81
0.50	1.22	1.22	−0.81	0.14	0.14	0.48	0.45	0.45	0.28	0.45	0.45	0.28	1.22	1.22	−0.81
0.75	1.22	1.22	−0.81	0.45	0.45	0.28	0.75	0.75	0.08	0.23	0.23	0.08	−1.09	−1.09	−0.66
1.00	−1.09	−1.09	−0.66	0.23	0.23	0.08	−0.05	−0.05	−0.09	0.07	0.07	0.11	0.89	0.89	0.53
1.25	0.89	0.89	0.53	0.07	0.07	0.11	0.29	0.29	0.24	0.29	0.29	0.24	0.89	0.89	0.53
1.50	0.89	0.89	0.53	0.29	0.29	0.24	0.51	0.51	0.37	0.47	0.47	0.37	−0.80	−0.80	0.43
1.75	−0.80	−0.80	0.43	0.47	0.47	0.37	0.27	0.27	0.48	0.27	0.27	0.48	−0.80	−0.80	0.43
2.00	−0.80	−0.80	0.43	0.27	0.27	0.48	0.07	0.07	0.59	0.07	0.07	0.39	−0.72	−0.72	−0.43

* Rounded to two decimal places
** Number of collisions in every timestep: 1st—02; 2nd—0; 3rd—01; 4th—02; 5th—0; 6th—01; 7th—0; 8th—01

Fig. B.11 Positions of the centre of mass of the particle at the 10th validation test

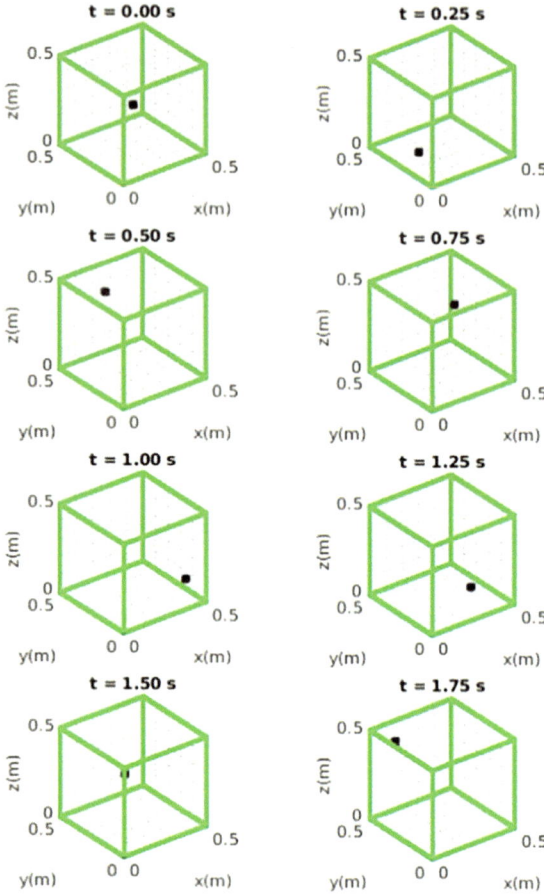

Table B.10 Results of the 10th validation test* (SI units)

t **	Input data									Output data					
	V_{ox}	V_{oy}	V_{oz}	C_{ox}	C_{oy}	C_{oz}	C_{lx}	C_{ly}	C_{lz}	C_{fx}	C_{fy}	C_{fz}	V_{fx}	V_{fy}	V_{fz}
0.25	−1.00	−1.50	−1.50	0.25	0.25	0.25	0.00	−0.12	−0.12	0.02	0.13	0.13	0.73	1.09	1.09
0.50	0.73	1.09	1.09	0.02	0.13	0.13	0.20	0.40	0.40	0.20	0.40	0.40	0.73	1.09	1.09
0.75	0.73	1.09	1.09	0.20	0.40	0.40	0.38	0.68	0.68	0.38	0.32	0.32	0.59	−0.89	−0.89
1.00	0.59	−0.89	−0.89	0.38	0.32	0.32	0.53	0.10	0.10	0.45	0.10	0.10	−0.53	−0.80	−0.80
1.25	−0.53	−0.80	−0.80	0.45	0.10	0.10	0.32	−0.10	−0.10	0.32	0.11	0.11	−0.43	0.65	0.65
1.50	−0.43	0.65	0.65	0.32	0.11	0.11	0.21	0.27	0.27	0.21	0.27	0.27	−0.43	0.65	0.65
1.75	−0.43	0.65	0.65	0.21	0.27	0.27	0.10	0.43	0.43	0.10	0.43	0.43	−0.43	0.65	0.65
2.00	−0.43	0.65	0.65	0.10	0.43	0.43	0.00	0.59	0.59	0.02	0.40	0.40	0.31	−0.47	−0.47

* Rounded to two decimal places

** Number of collisions in every timestep: 1st—02; 2nd—0; 3rd—01; 4th—01; 5th—01; 6th—0; 7th—0; 8th—02

Fig. B.12 Positions of the centre of mass of the particle at the 11th validation test

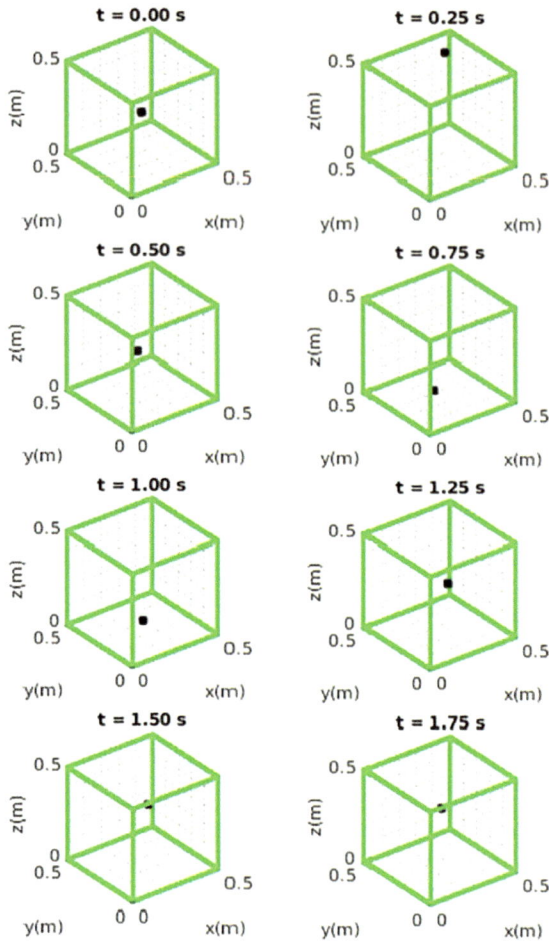

Table B.11 Results of the 11th validation test* (SI units)

t **	Input data									Output data					
	V_{ox}	V_{oy}	V_{oz}	C_{ox}	C_{oy}	C_{oz}	C_{1x}	C_{1y}	C_{1z}	C_{fx}	C_{fy}	C_{fz}	V_{fx}	V_{fy}	V_{fz}
0.25	1.50	1.50	1.00	0.25	0.25	0.25	0.62	0.62	0.50	0.38	0.38	0.48	−0.97	−0.97	−0.65
0.50	0.38	0.38	0.48	−0.97	−0.97	−0.65	0.14	0.14	0.32	0.14	0.14	0.32	−0.97	−0.97	−0.65
0.75	−0.97	−0.97	−0.65	0.14	0.14	0.32	−0.10	−0.10	0.16	0.10	0.10	0.16	0.70	0.70	−0.52
1.00	0.70	0.70	−0.52	0.10	0.10	0.16	0.28	0.28	0.03	0.28	0.28	0.03	0.70	0.70	−0.52
1.25	0.70	0.70	−0.52	0.28	0.28	0.03	0.45	0.45	−0.10	0.45	0.45	0.10	0.63	0.63	0.42
1.50	0.63	0.63	0.42	0.45	0.45	0.10	0.61	0.61	0.21	0.40	0.40	0.21	−0.45	−0.45	0.34
1.75	−0.45	−0.45	0.34	0.40	0.40	0.21	0.28	0.28	0.29	0.28	0.28	0.29	−0.45	−0.45	0.34
2.00	−0.45	−0.45	0.34	0.28	0.28	0.29	0.17	0.17	0.38	0.17	0.17	0.38	−0.45	−0.45	0.34

* Rounded to two decimal places
** Number of collisions in every timestep: 1st—02; 2nd—0; 3rd—01; 4th—0; 5th—01; 6th—01; 7th—0; 8th—0

Fig. B.13 Positions of the
centre of mass of the particle at
the 12th validation test

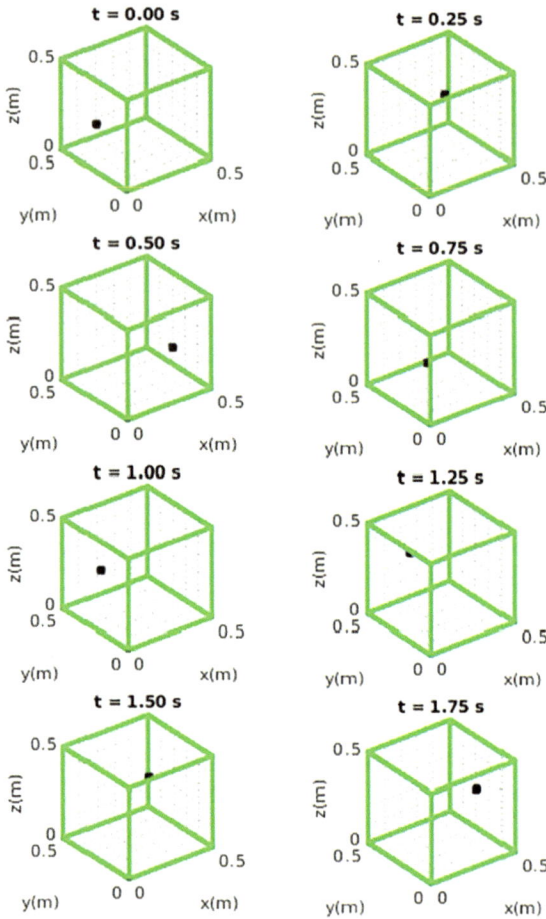

Table B.12 Results of the 12th validation test* (SI units)

t **	Input data									Output data					
	V_{ox}	V_{oy}	V_{oz}	C_{ox}	C_{oy}	C_{oz}	C_{1x}	C_{1y}	C_{1z}	C_{fx}	C_{fy}	C_{fz}	V_{fx}	V_{fy}	V_{fz}
0.25	1.50	1.50	0.00	0.01	0.25	0.25	0.39	0.62	0.25	0.39	0.38	0.25	1.20	−1.20	0.00
0.50	1.20	−1.20	0.00	0.39	0.38	0.25	0.69	0.08	0.25	0.33	0.08	0.25	−0.96	−0.96	0.00
0.75	−0.96	−0.96	0.00	0.33	0.08	0.25	0.09	−0.16	0.25	0.09	0.14	0.25	−0.77	0.77	0.00
1.00	−0.77	0.77	0.00	0.09	0.14	0.25	−0.10	0.34	0.25	0.10	0.34	0.25	0.61	0.61	0.00
1.25	0.61	0.61	0.00	0.10	0.34	0.25	0.25	0.49	0.25	0.25	0.49	0.25	0.61	0.61	0.00
1.50	0.61	0.61	0.00	0.25	0.49	0.25	0.40	0.64	0.25	0.40	0.37	0.25	0.49	−0.49	0.00
1.75	0.49	−0.49	0.00	0.40	0.37	0.25	0.53	0.24	0.25	0.46	0.24	0.25	−0.39	−0.39	0.00
2.00	−0.39	−0.39	0.00	0.46	0.24	0.25	0.36	0.15	0.25	0.36	0.15	0.25	−0.39	−0.39	0.00

* Rounded to two decimal places
** Number of collisions in every timestep: 1st—01; 2nd—01; 3rd—01; 4th—01; 5th—0; 6th—01; 7th—01; 8th—0

Fig. B.14 Positions of the centre of mass of the particle at the 13th validation test

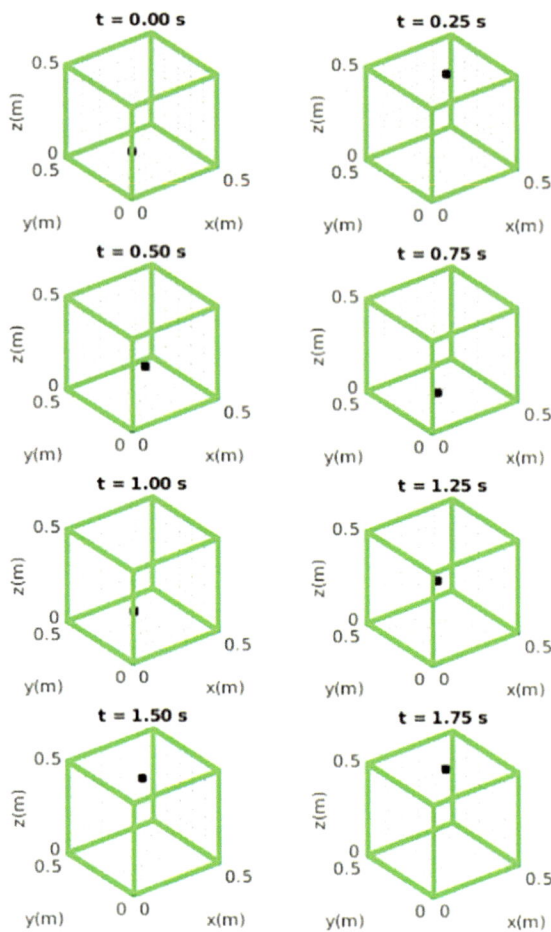

Table B.13 Results of the 13th validation test* (SI units)

t**	Input data									Output data					
	V_{ox}	V_{oy}	V_{oz}	C_{ox}	C_{oy}	C_{oz}	C_{1x}	C_{1y}	C_{1z}	C_{fx}	C_{fy}	C_{fz}	V_{fx}	V_{fy}	V_{fz}
0.25	1.50	1.50	1.50	0.01	0.01	0.25	0.39	0.39	0.62	0.39	0.39	0.38	1.20	1.20	−1.20
0.50	1.20	1.20	−1.20	0.39	0.39	0.38	0.69	0.69	0.08	0.33	0.33	0.08	−0.77	−0.77	−0.77
0.75	−0.77	−0.77	−0.77	0.33	0.33	0.08	0.14	0.14	−0.11	0.14	0.14	0.11	−0.61	−0.61	0.61
1.00	−0.61	−0.61	0.61	0.14	0.14	0.11	−0.01	−0.01	0.26	0.03	0.03	0.26	0.39	0.39	0.39
1.25	0.39	0.39	0.39	0.03	0.03	0.26	0.13	0.13	0.36	0.13	0.13	0.36	0.39	0.39	0.39
1.50	0.39	0.39	0.39	0.13	0.13	0.36	0.22	0.22	0.46	0.22	0.22	0.46	0.39	0.39	0.39
1.75	0.39	0.39	0.39	0.22	0.22	0.46	0.32	0.32	0.55	0.32	0.32	0.44	0.31	0.31	−0.31
2.00	0.31	0.31	−0.31	0.32	0.32	0.44	0.40	0.40	0.36	0.40	0.40	0.36	0.31	0.31	−0.31

* Rounded to two decimal places
** Number of collisions in every timestep: 1st—01; 2nd—01; 3rd—01; 4th—01; 5th—0; 6th—0; 7th—01; 8th—0

Fig. B.15 Positions of the centre of mass of the particle at the 14th validation test

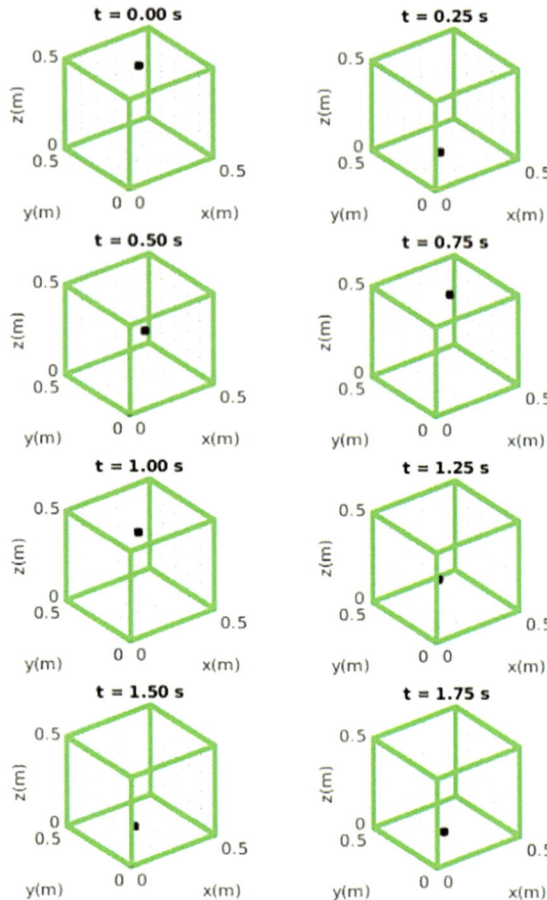

Table B.14 Results of the 14th validation test* (SI units)

t **	Input data									Output data					
	V_{ox}	V_{oy}	V_{oz}	C_{ox}	C_{oy}	C_{oz}	C_{1x}	C_{1y}	C_{1z}	C_{fx}	C_{fy}	C_{fz}	V_{fx}	V_{fy}	V_{fz}
0.25	−1.50	−1.50	−1.50	0.25	0.25	0.49	−0.12	−0.12	0.11	0.12	0.12	0.11	1.08	1.08	−1.08
0.50	1.08	1.08	−1.08	0.12	0.12	0.11	0.40	0.40	−0.16	0.40	0.40	0.15	0.92	0.92	0.92
0.75	0.92	0.92	0.92	0.40	0.40	0.15	0.63	0.63	0.38	0.37	0.37	0.38	−0.67	−0.67	0.67
1.00	−0.67	−0.67	0.67	0.37	0.37	0.38	0.21	0.21	0.55	0.21	0.21	0.44	−0.57	−0.57	−0.57
1.25	−0.57	−0.57	−0.57	0.21	0.21	0.44	0.07	0.07	0.30	0.07	0.07	0.30	−0.57	−0.57	−0.57
1.50	−0.57	−0.57	−0.57	0.07	0.07	0.30	−0.07	−0.07	0.16	0.08	0.08	0.16	0.41	0.41	−0.41
1.75	0.41	0.41	−0.41	0.08	0.08	0.16	0.18	0.18	0.06	0.18	0.18	0.06	0.41	0.41	−0.41
2.00	0.41	0.41	−0.41	0.18	0.18	0.06	0.29	0.29	−0.05	0.29	0.29	0.06	0.35	0.35	0.35

* Rounded to two decimal places
** Number of collisions in every timestep: 1st—01; 2nd—01; 3rd—01; 4th—01; 5th—0; 6th—01; 7th—0; 8th—01

Fig. B.16 Positions of the
centre of mass of the particle at
the 15th validation test

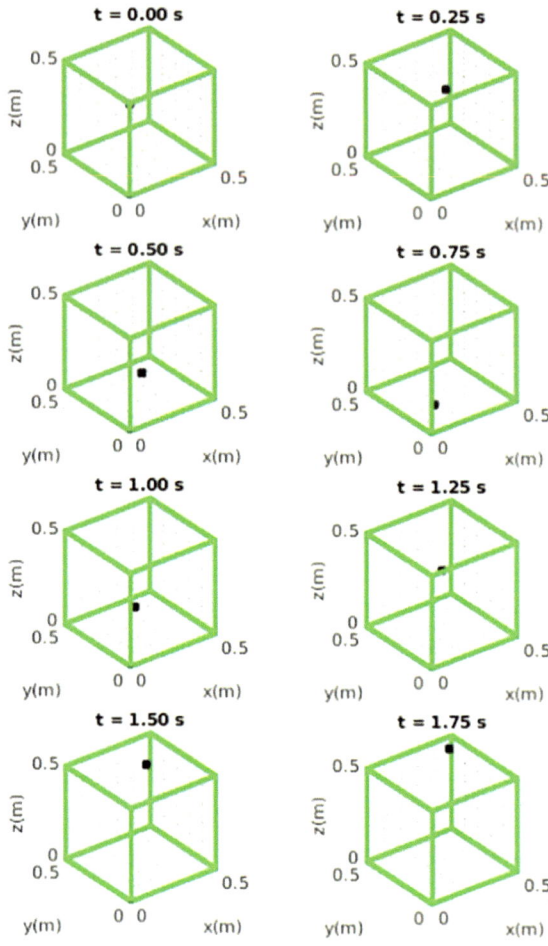

Table B.15 Results of the 15th validation test* (SI units)

t**	Input data									Output data					
	V_{ox}	V_{oy}	V_{oz}	C_{ox}	C_{oy}	C_{oz}	C_{1x}	C_{1y}	C_{1z}	C_{fx}	C_{fy}	C_{fz}	V_{fx}	V_{fy}	V_{fz}
0.25	1.50	1.50	1.00	0.01	0.01	0.49	0.39	0.39	0.74	0.39	0.39	0.28	1.27	1.27	−0.85
0.50	1.27	1.27	−0.85	0.39	0.39	0.28	0.70	0.70	0.07	0.31	0.31	0.07	−0.92	−0.92	−0.61
0.75	−0.92	−0.92	−0.61	0.31	0.31	0.07	0.08	0.08	−0.09	0.08	0.08	0.09	−0.78	−0.78	0.52
1.00	−0.78	−0.78	0.52	0.08	0.08	0.09	−0.12	−0.12	0.22	0.12	0.12	0.22	0.57	0.57	0.38
1.25	0.57	0.57	0.38	0.12	0.12	0.22	0.26	0.26	0.32	0.26	0.26	0.32	0.57	0.57	0.38
1.50	0.57	0.57	0.38	0.26	0.26	0.32	0.40	0.40	0.41	0.40	0.40	0.41	0.57	0.57	0.38
1.75	0.57	0.57	0.38	0.40	0.40	0.41	0.54	0.54	0.51	0.45	0.45	0.48	−0.35	−0.35	−0.23
2.00	−0.35	−0.35	−0.23	0.45	0.45	0.48	0.36	0.36	0.42	0.36	0.36	0.42	−0.35	−0.35	−0.23

* Rounded to two decimal places
** Number of collisions in every timestep: 1st—01; 2nd—01; 3rd—01; 4th—01; 5th—0; 6th—0; 7th—02; 8th—0

Fig. B.17 Positions of the
centre of mass of the particle at
the 16th validation test

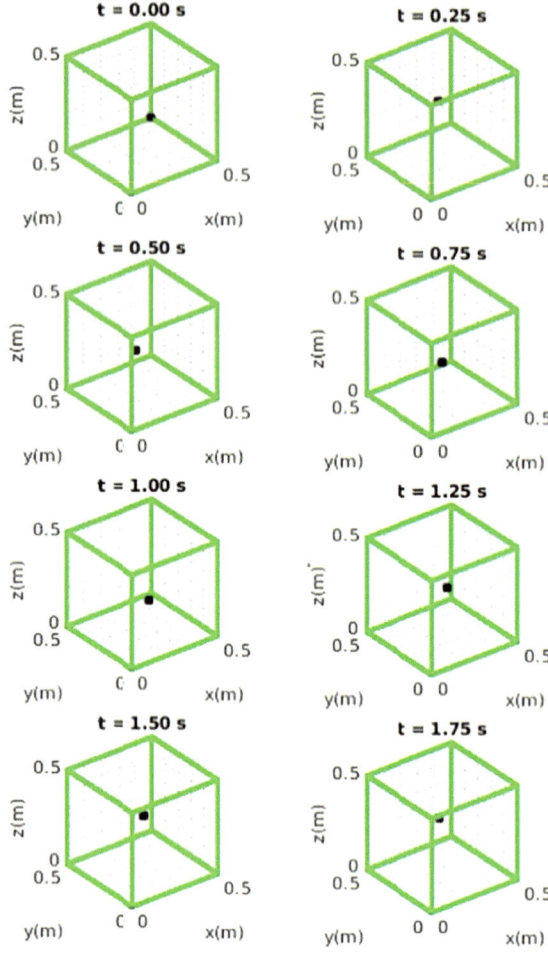

Table B.16 Results of the 16th validation test* (SI units)

t**	Input data									Output data					
	V_{ox}	V_{oy}	V_{oz}	C_{ox}	C_{oy}	C_{oz}	C_{1x}	C_{1y}	C_{1z}	C_{fx}	C_{fy}	C_{fz}	V_{fx}	V_{fy}	V_{fz}
0.25	1.50	1.50	1.50	0.49	0.49	0.01	0.86	0.86	0.39	0.17	0.17	0.39	−1.08	−1.08	−1.08
0.50	−1.08	−1.08	−1.08	0.17	0.17	0.39	−0.10	−0.10	0.66	0.10	0.10	0.35	0.67	0.67	−0.67
0.75	0.67	0.67	−0.67	0.10	0.10	0.35	0.27	0.27	0.18	0.27	0.27	0.18	0.67	0.670	−0.67
1.00	0.67	0.670	−0.67	0.27	0.27	0.18	0.44	0.44	0.02	0.44	0.44	0.02	0.67	0.67	−0.67
1.25	0.67	0.67	−0.67	0.44	0.44	0.02	0.60	0.60	−0.15	0.39	0.39	0.15	−0.41	−0.41	0.41
1.50	−0.41	−0.41	0.41	0.39	0.39	0.15	0.29	0.29	0.25	0.29	0.29	0.25	−0.41	−0.41	0.41
1.75	−0.41	−0.41	0.41	0.29	0.29	0.25	0.19	0.19	0.35	0.19	0.19	0.35	−0.41	−0.41	0.41
2.00	−0.41	−0.41	0.41	0.19	0.19	0.35	0.09	0.09	0.45	0.09	0.09	0.45	−0.41	−0.41	0.41

* Rounded to two decimal places

** Number of collisions in every timestep: 1st—01; 2nd—02; 3rd—0; 4th—0; 5th—02; 6th—0; 7th—0; 8th—0

Index

© The Editor(s) (if applicable) and The Author(s), under exclusive license   97
to Springer Nature Switzerland AG 2025
C. A. D. Fraga Filho, *Reflective Boundary Conditions in SPH Fluid Dynamics
Simulation*, Synthesis Lectures on Mechanical Engineering,
https://doi.org/10.1007/978-3-031-71582-2